Conservation of Genetic Resources

Springer
Berlin
Heidelberg
New York
Barcelona
Hong Kong
London
Milan
Paris
Singapore
Tokyo

D. Virchow

Conservation of Genetic Resources

Costs and Implications for a Sustainable Utilization of Plant Genetic Resources for Food and Agriculture

With 28 Figures and 31 Tables

Dr. DETLEF VIRCHOW
ZEF Bonn – Zentrum für Entwicklungsforschung
(Center for Development Research)
Universität Bonn
Walter-Flex-Straße 3
53113 Bonn, Germany

Printed as dissertation with approval of the Agricultural Faculty of the Christian-Albrechts-Universität of Kiel

---

ISBN 3-540-65343-0 Springer-Verlag Berlin Heidelberg New York

Die Deutsche Bibliothek – CIP-Einheitsaufnahme
Virchow, Detlef: Conservation of genetic resources: costs and implications for a sustainable utilization of plant genetic resources for food and agriculture; with 31 tables / Detlef Virchow. – Berlin; Heidelberg; New York; Barcelona; Hong Kong; London; Milan; Paris; Singapore; Tokyo: Springer, 1999
Zugl.: Kiel, Univ., Diss., 1998
ISBN 3-540-65343-0

Production: PRO EDIT GmbH, D-69126 Heidelberg
Cover design: de'blik, D-10435 Berlin
Typesetting: Camera-ready from the author

SPIN 10696497 30/3136-5 4 3 2 1 0 – Printed on acid-free paper

Acknowledgements

This research was made possible through a grant by the *Vater und Sohn Eiselen-Stiftung, Ulm*. The generous and far-sighted decision by the Foundation and Senator Dr. Hermann Eiselen to establish this research in an area which will be critical for world food security in the long run is gratefully acknowledged. This grant also made it possible for me to become directly involved as a member of the FAO Secretariat with the "International Technical Conference on Plant Genetic Resources" in Leipzig, 1996 and FAO leadership in this important event.

In the course of this research project I benefited from the helpful conversation, advice, and cooperation of many people, in particular including:

at FAO, Prof. Dr. Hartwig de Haen, Assistant Director-General; Prof. Dr. Jose T. Esquinas-Alcazar, Dr. Umberto Menini, Dr. Murthi Anishetty, Dr. Clive Stannard, as well as the FAO project team coordinated by Dr. Cary Fowler and Dr. David Cooper and consisting among others of Suzanne Sharrock, Kristin Kolshus, Iqbal Kermali; and at IPGRI, Dr. Jan Engels and others;

in India, Prof. Dr. R.B. Singh, Director of the Indian Agricultural Research Institute; Dr. N.C. Jain, IARI; Dr. K.P.S. Chandel, Director of the National Bureau of Plant Genetic Resources; Prof. Dr. Kumar, Head of Division of Agricultural Economics, IARC; Prof. Dr. Anil K. Gupta, Indian Institute of Management; Prof. Dr. M.S. Swaminathan, M. Geetha Rani, both M.S. Swaminathan Research Foundation; Dr. Eva Weltzien-Rattunde, Dr. N.K. Rao, Dr. A.K. Singh, all ICRISAT;

in the German PGR system, Dr. Wilbert Himmighofen, German Federal Ministry of Food, Agriculture and Forestry; Dr. Frank Begemann and Dr. Anja Oetmann, both Center for Information on Genetic Resources; Dr. Lothar Frese and Stefan Bücken, both Braunschweig Genetic Resources Center; Prof. Dr. Karl Hammer and Dr. Helmut Gäde, both Institute of Crop Science and Plant Breeding;

at Kiel, my colleagues, especially Dr. Peter Wehrheim, Katinka Weinberger and Matin Qaim.

Many others also provided valuable inputs, comments and help, especially Regina Schulz-Giese and Derek Z. Gill.

Finally, I am particularly indebted to the pleasant and very efficient cooperation with my supervisor Prof. Dr. Joachim von Braun whose intellectual guidance and support are most gratefully acknowledged.

Although the present text was prepared with the assistance of the above-mentioned persons whose collaboration is greatly appreciated, I alone am responsible for any errors, inaccuracies or omissions.

Detlef Virchow

Dedicated to
Christopher, Stephen, Laurenz and Lioba
who represent the generation to come,
and whose future needs should influence
the present generation's action.

Contents

Tables

Figures

Appendices

List of Acronyms and Abbreviations

aMSS	areal Minimum Safety Standard
BAZ	Federal Center for Breeding Research on Cultivated Plants Bundesanstalt für Züchtungsforschung an Kulturpflanzen
BfN	Federal Agency for Nature Conservation Bundesamt für Naturschutz
BGCI	Botanic Gardens Conservation International
BGRC	Braunschweig Genetic Resources Center
BMBF	German Federal Ministry of Education, Science, Research and Technology Bundesministerium für Bildung, Wissenschaft, Forschung und Technologie
BMJ	German Federal Ministry of Justice Bundesministerium der Justiz
BML	German Federal Ministry of Food, Agriculture and Forestry Bundesministerium für Ernährung, Landwirtschaft und Forsten
BMU	German Federal Ministry of the Environment, Nature Conservation and Nuclear Safety Bundesministerium für Umwelt, Naturschutz und Reaktorsicherheit
BMZ	German Federal Ministry of Economic Cooperation and Development Bundeministerium für wirtschaftliche Zusammenarbeit und Entwicklung
BMWi	German Federal Ministry of Economics Bundeministerium für Wirtschaft
CBD	Convention on Biological Diversity
CDG	Carl-Duisberg-Association Carl-Duisberg-Gesellschaft
CGIAR	Consultative Group on International Agricultural Research
CIAT	International Center for Tropical Agriculture, CGIAR
CIMMYT	Centro Internacional de Mejoramiento de Maís y Trigo, International Maize and Wheat Center - CGIAR
CIP	Centro Internacional de la Papa, International Potato Center - CGIAR
CITES	Convention on the International Trade in Endangered Species of Wild Fauna and Flora

CoP/CBD	Conference of the Parties of the Convention on Biological Diversity
CGRFA	Commission on Genetic Resources for Food and Agriculture, FAO
CSC	Conservation and Service Center for PGRFA Conservation
DAAD	German Academic Exchange Service Deutscher Akademischer Austauschdienst
DSE	German Foundation for International Development Deutsche Stiftung für Internationale Entwicklung
EU	European Union
FAL	Federal Research Center of Agriculture Bundesforschungsanstalt für Landwirtschaft
FAO	Food and Agriculture Organisation
GATT	General Agreement on Tariffs and Trade
GCF	Genetically Coded Function
GCI	Genetically Coded Information
GDP	Gross Domestic Product
GEF	Global Environment Facility
GRAIN	Genetic Resources Action International
GTZ	German Agency for Technical Cooperation Gesellschaft für Technische Zusammenarbeit
HYV	high-yielding variety
IARCs	International Agricultural Research Centers
ICARDA	International Center for Agricultural Research in the Dry Areas - CGIAR
ICRAF	International Center for Research in Agroforestry - CGIAR
ICRISAT	International Crops Research Institute for the Semi-Arid Tropics - CGIAR
IFAD	International Fund for Agricultural Development
IGR	Center for Information on Genetic Resources at ZADI Informationszentrum für Genetische Ressourcen
IITA	International Institute of Tropical Agriculture - CGIAR
INIBAP	International Network for the Improvement of Bananas and Plantains - CGIAR
IPGRI	International Plant Genetic Resources Institute - CGIAR
IPK	Institute of Plant Genetics and Crop Plant Research Institut für Pflanzengenetik und Kulturpflanzenforschung Germany
IPR	Intellectual Property Rights
IRRI	International Rice Research Institute - CGIAR
ISAAA	International Service for the Acquisition of Agri-biotech Applications
ITCPGR	International Technical Conference on Plant Gentic Resources
IUCN	The World Conservation Union Gland
IUPGR	International Undertaking on Plant Genetic Resources
MAB	Man and the Biosphere Programme, UNESCO

MPI	Max-Planck-Institute Max-Planck-Institut
MSS	Minimum Safety Standard
NARS	National Agricultural Research Systems
NBPGR	National Bureau of Plant Genetic Resources, India
NGB	Nordic Gene Bank, Sweden
NGO	Non-Governmental Organisation
OECD	Organisation for Economic Cooperation and Development Paris
ODA	Official Development Assistance
PGR	Plant Genetic Resources
PGRFA	Plant Genetic Resources for Food and Agriculture
RAFI	Rural Advancement Foundation International
SGRP	System-wide Genetic Resources Programme - CGIAR
SPGRC	Southern African Development Community-Plant Genetic Resources Center
TRIPS	Trade-Related Aspects of Intellectual Property Rights
UK	United Kingdom
UN	United Nations
UNCED	United Nations Conference on Environment and Development ("Rio Conference")
UNDP	United Nations Development Programme
UNEP	United Nations Environment Programme
UNESCO	United Nations Educational, Scientific and Cultural Organisation
UPOV	Union for the Protection of New Varieties of Plants Union pour la Protection des Obtentions Végétales
WIEWS	World Information and Early Warning System, FAO
WRI	World Resources Institute
WTO	World Trade Organisation
WWF	World Wide Fund for Nature Gland
ZADI	Center for Agricultural Documentation and Information Zentralstelle für Agrardokumentation und –information

1 Introduction

During the past decade the importance of natural resources for a sustainable agricultural development has been increasingly discussed at international forums and conferences. The degradation of soil, the climatic change, and the increasing shortage of water were identified as the main obstacles for a sustainable agricultural development. Even though the institutional process for awareness, conservation and exchange of genetic resources started in the 1950s (see Chap. 4), the negotiation and signing of the Convention on Biological Diversity (CBD) at the *United Nations Conference on Environment and Development* in Rio de Janeiro took up to 1992. The most important international conference for the specific issue of agrobiodiversity was held in Leipzig in 1996, where the first report on the state of the world's plant genetic resources for food and agriculture (PGRFA) was presented and the first global plan of action for the conservation of PGRFA was adopted (FAO, 1996c). Aside from the sustainable management of soil, water, and air, it now seems to be accepted that the sustainable management of genetic resources is one of the four indispensable preconditions for a sustainable agriculture. Genetic resources are necessary for the breeding efforts to solve the future challenges. Nevertheless, it has been only recently that agricultural economists are beginning to direct research at the importance of plant genetic resources conservation and utilization due to its considerable long run impact on agricultural development and food security (von Braun, 1994a).

1.1 Problems

At the same time, although the importance of biodiversity for a sustainable agriculture is now commonly recognized, the genetic diversity in farmers' fields is reduced through the displacement of traditional varieties by modern varieties and introduced crops[1]. Because a growing share of food is provided by a limited number of crops and varieties of each crop species, it is crucial not only to conserve the existing diversity, but also to use its components in a sustainable way to meet present and future needs. As demand for genetic resources increases because of new technology and new applications of biotechnology in agriculture, in pharmaceutics, in environmental engineering, and in the development of new

[1] On definitions see Chap. 1.2

materials the conservation of genetic resources is becoming increasingly important.

Increasing production - and more recently stabilizing production at a high level – have been a question of increasing inputs like fertilizer and pesticides, and improving management systems, mainly irrigation and crop varieties. Improving varieties has been seen merely as a function depending on the variables of allocation of human and financial resources to research and development, i.e., breeding. The importance of plant genetic resources derived from local varieties or from wild relatives of domesticated crops has not been reflected economically, because the free and unlimited availability of plant genetic resources has been taken for granted and the small but important input of traditional varieties or wild crop relatives was often neglected in breeding programs.

Breeding - and therewith PGRFA, as a small but essential input for breeding programs - will become an even more crucial factor in the attempt at reducing the number of hungry people in the world by half by the year 2015, as was called for by the World Food Summit 1996 (FAO, 1996e). The growing world population must be fed from an almost constant land and water resource base (Heidhues, 1994). Between 1962 and 1989 the combined total food consumption of 90 developing countries rose by 134%[2] (Boongarts, 1996). In those years, agricultural production growth based on the expansion of area increase and the intensification of area through improved management and breeding. One of the biggest challenges facing the world today is keeping up with the increasing food demand because of population and income growth. In this context, food production will require a sustainable utilization of natural resources like land and water and it will demand a more intensive and intelligent utilization of genetic resources. Short-term projections for 2010 or 2020 based on econometric models (Mitchell et al., 1995; Alexandratos, 1995; Agcaoili et al., 1995) foresee no major obstacles impeding the continued expansion of the food supply, given a continued increase of yields, in addition to other factors facilitating the dissemination of new technologies [3]. But because of the evidence of new yield ceilings under given circumstances, which have been approached by producers in some countries (Alexandratos, 1995), increasing food production by breeding depends upon the continued access to and the incorporation of genetic resources into the breeding programs.

Since the beginning of agriculture, farmers have been using and improving plant genetic resources by domesticating, cultivating, and improving crop varieties. In addition, they have been storing seeds for the next planting season. Storing plant genetic resources for the sake of conserving varieties threatened by

---

[2] This was achieved mainly by the increases of cultivated land, cropping frequency, and crop yield. Whereby the increase in average crop yield was the most important change, the yield increase over that period of time is estimated to be 41 to 75%, depending on the approach (either taking the average of each country (largest countries like India and China are influencing the outcome more) or taking the average of the changes of all countries together.

[3] Improvement of economic policies, investment in research and human resources are examples for other factors facilitating the dissemination of new technologies.

extinction, however, entered into the scientific discussion through the worldwide systematic survey, collection and conservation activities of plant genetic resources undertaken by N. I. Vavilov since the 1930s. The modern conservation movement started with endangered animal species (whales, elephants, rhinos etc.). However, an awareness of the importance of all species emerged and it was recognized that biodiversity not only benefits the sustainability of the ecosystems itself but also creates opportunities for people in a wide range of sectors (i.e., agriculture, tourism, research, health, disaster prevention; McNeely, Jeffrey, 1996).

Since sensitivity to the irreversible loss of genetic resources for food and agriculture has emerged, immense energy, human capacity, and financial resources have been spent for the collection and conservation of plant genetic resources and the establishment of an institutional framework on an international, national, and local level. There are 6.2 million accessions of 80 different crops stored in 1,320 genebanks and related facilities in 131 countries (FAO, 1996a).

At present, the PGRFA conservation efforts seem to lack an overall system: even though the loss of genetic diversity in agriculturally relevant plants is evident and conservation activities are taking place, the knowledge concerning the rate of extinction is not indisputable and the marginal benefit of the conserved varieties is open to wide speculations. Some estimate the marginal benefit of each crop variety to be over 2,000 US $ a year (Evenson, 1996a), while others are concerned about conserving redundant genetic resources (Simpson and Sedjo, 1996). Consequently, the investments in the conservation of PGRFA are undertaken without clear optimality criteria. This, again, relates to the uncertainty of expected benefits.

A conservation strategy utilizing ex situ conservation (e.g., genebanks) as well as in situ conservation possibilities (genetic resources conservation or management in farmers' fields) is emerging only recently. The combination of ex situ and in situ conservation methods seems indispensable, not only because a loss of plant genetic resources is becoming apparent in farmers' fields, but also because ex situ conservation methods, e.g., genebanks, are not capable of preserving all plant genetic resources in their original condition without the loss of parts or entire collections. Hence, only a well-balanced combination of ex situ and in situ conservation seems to meet all the relevant objectives.

Furthermore, because of the rapidly increasing awareness of the importance of PGRFA and their silent reduction, conservation activities were intensified, but seldom managed accurately and efficiently. Consequently, the shelves of the existing ex situ facilities are overloaded but the information on the accessions is often poorly documented (FAO, 1996a)[4]. Therefore main information, e.g., how many and what kind of varieties are conserved ex situ, is lacking. Additionally, PGRFA conservation has recently not only been seen as freezing accessions for the generations to come, but mainly as management of information combined with service functions for all those demanding PGRFA. Without a global, national and local survey and inventory of existing crop species and varieties, there is no

---

[4] The expression for accessions without or with very little information is establishing itself as 'UFOs': unknown frozen objects.

scientific means by which to decide how many and which crop species must be conserved on a global level and in the interest of intergenerational equity.

The above-mentioned deficits of information and uncertainties are hindering an economically efficient approach to optimizing agrobiodiversity conservation. Because of lacking estimations on:

- the value of PGRFA for global welfare (e.g., value of PGRFA for breeding), or the cost of their extinction,
- the rate of PGRFA extinction, and
- the costs of conservation,

investments in PGRFA conservation are most likely sub-optimal at the margin. Additionally, allocative problems such as the imbalance between the shared costs and the benefits of conservation hamper an optimal conservation at all levels. For example, some countries with a high amount of unique PGRFA belong to the poorest countries in the world, where investment in conservation is constrained by very limited resources and other priorities for the use of available funds (von Braun and Virchow, 1997).

Despite the existing uncertainties concerning the economic value of PGRFA for national and global welfare, the political will, expressed by all governments present at the Technical Conference on Plant Genetic Resources for Food and Agriculture in Leipzig in 1996, stressed the importance of genetic resources conservation (FAO, 1996b). This lent support to continued conservation of PGRFA, even though long-term conservation activities face strong competition from other often more short-term development activities for the allocation of financial resources.

Considering these circumstances, there is a need for cost-effective and efficient strategies for PGRFA conservation, in addition to further scientific and economic research. Cost-efficient conservation will reduce the risk of losing unique genetically coded information and reduce the problem of allocating too extensive financial resources to conservation activities. Although the political discussion is focused on the issue of "fair and equitable sharing" of the benefits derived from the use of PGRFA, and the economic discussion is focused on the value of plant genetic resources and in situ conservation, an intensive analysis of the costs of conservation activities has been neglected. Therefore this study will focus on the costs of conservation and the efficiency and effectiveness of the conservation strategies.

1.2 Definitions of Genetic Resources

Preservation, conservation:

There are two fundamentally different concepts of protecting nature: **preservation** implies the protection of areas, species or other units of nature from any kind of human activities or influence, whereby **conservation** implies the protection of parts of nature for anthropogenic sustainable utilization. Consequently, conservation policies may include preservation activities to safeguard specific endemic species for future utilization.

Biodiversity:

The term **biodiversity** was introduced into the scientific discussion by Wilson (1988) by shortening the common term biological diversity. Nevertheless, he did not define biodiversity. According to Banham (1993), biodiversity is the variety of all living organisms at all levels and is divided into three categories:

- genetic diversity - the genetic variation within and between one population (all possible gene combinations within one species);
- species diversity - the variation between species of one region;
- diversity of ecosystems - the variation of components of a given ecosystem, providing the necessary conditions for populations of species.

In the broader discussion, biodiversity also represents a diversity of values. The preamble of the UN Convention on Biological Diversity states the three types of variations mentioned above and refers to the *"... intrinsic value of biological diversity and of the ecological, genetic, social, economic, scientific, educational, cultural, recreational and aesthetic values of biological diversity and its components ..."* (Preamble of UNEP, 1994a).

Agrobiodiversity and PGRFA:

Agrobiodiversity is defined broadly as *"... that part of biodiversity which nurtures people and which are nurtured by people ..."* (FAO, 1995a, paragraph 67). For reasons of functionality, agrobiodiversity is defined here as the diversity of existing domesticated plants and animals and is categorized here for agricultural crops as follows[5]:

- genetic diversity - the diversity of genetic variation in one variety;
- varietal diversity - the diversity of varieties in one crop species;
- species diversity - the diversity of crop species in one region.

In general, the term diversity has no operational value for analyzing, valuing, and devising efficient conservation options on the basis of economic instruments.

Plant genetic resources for food and agriculture (PGRFA) is the general expression for the material growing in farmers' fields and their wild crop relatives, as well as material which is conserved, exchanged, utilized - and threatened. PGRFA as a distinct part of the general plant genetic resources includes resources contributing to people's livelihoods by providing food, medicine, feed for domestic animals, fiber, clothing, shelter and energy. PGRFA are the inputs for breeding as well as for biotechnology, including genetic engineering.

---

[5] In this study, agrobiodiversity will be used for crop diversity only.

Species, varieties, cultivars:

A **species** is defined as a population in which a free gene flow and exchange exists and the individuals of the population are able to produce descendant. Species relevant for food and agriculture are conventionally divided into cultivated **varieties**.

Modern varieties are - according to the seed certification and plant breeders' rights - characterized by distinctness, uniformity and stability (UPOV, 1992). These varieties have been subject to breeding efforts and are protected in various protection acts and registered under some seed trade acts. Registered varieties, for which the protection period has expired and which may no longer be traded, are listed in the country's seed office under the term **old cultivars**. Varieties grown locally by farmers are **landraces**, which have been adapted to certain environmental conditions and are not registered in official seed variety lists. They are derived from various sources. In general, landraces contain high levels of genetic diversity (e.g., Alika et al., 1993, Ceccarelli et al., 1992) and exceed the modern varieties by numbers (Brush, 1991; Bolster, 1985)[6]. Therefore, landraces should be defined as populations and not as varieties. Consequently landraces are the main focus of conservation efforts. Instead of landraces, developing countries - for instance Ethiopia - prefer to call these populations "farmers' varieties", crediting the breeding work done by farmers for centuries.

Plant breeders normally use **advanced cultivars** or **breeders' lines** as base for further breeding activities or for creating a new variety. This material is derived from elder varieties, old cultivars or landraces, and is improved by breeding activities. The base material for a breeders' line may also be **wild or weedy crop relatives**. These are distant relatives of crop species and coexist with the agricultural crops used and are often major weeds in the field of domesticated crops. Wild or weedy crop relatives exist in or next to farmers' fields as well as in the natural vegetation.

Accession:

An **accession** is the planting material of a variety stored in a conservation facility. An accession represents the smallest storable unit of a cropvariety. By cereals, an accession consists of approximately 500 to 1,000 seeds, which are dried and usually conserved cold or frozen (Hammer, 1995).

[6] There are communities in the Andes which grow 178 locally named potato varieties (Brush, 1991), or in the Peruvian Amazon 61 distinct cultivars of cassava are grown (Bolster, 1985).

1.3 Research Questions and Approach

1.3.1 Research Questions

The international community, governments, breeding companies, and farmers have become aware of the persisting threat of losing plant genetic resources. They are willing to conserve PGRFA, but the conservation strategies and the institutional mechanisms including transaction mechanisms for PGRFA as well as financial mechanisms have not yet been agreed on. As basis for collection and conservation activities, the institutional framework has changed dramatically since the signing of the Convention on Biological Diversity in 1992. The first systematic attempt to develop a comprehensive, co-ordinated plan to conserve PGRFA was not made until 1996. The "Leipzig Declaration on Conservation and Sustainable Utilization of Plant Genetic Resources for Food and Agriculture" and the "Global Plan of Action for the Conservation and Sustainable Utilization of Plant Genetic Resources for Food and Agriculture" were adopted by 150 countries in Leipzig, 1996 (FAO, 1996b), signalizing the political will to continue and to increase the effort to conserve PGRFA.

This study will contribute to the development-policy discussion by raising and discussing some institutional economics research questions as well as analyzing the costs arising by conserving PGRFA. The research questions raised include:

1. *To what extent is genetic loss relevant to agriculture?* It is necessary to collect information on the order of magnitude of genetic losses before any decision to invest in conservation activities can be undertaken. Are genetic losses in agriculturally relevant plants already threatening the breeding activities of the future?
2. *What are the main determinants of genetic losses?* It is important to understand the causes and underlying determinants of the loss for any conservation strategy and planning so as to react by taking this information into account. The Green Revolution is often blamed for genetic extinction, and on the eve of a new, genetically engineered revolution, the perceived threat of significant and severe genetic loss is widespread. It is essential to understand the interrelation between development and genetic extinction.
3. *What are the relevant differences between general biodiversity and diversity in agricultural crops?* A lot of work is being done on the conservation of biodiversity in general, neglecting agriculturally relevant genetic resources or suggesting them implicitly. Aspects of diversity, conservation, and utilization must be differentiated.
4. *Which methods of PGRFA conservation are practiced and what are their complementarities?* An overall systems perspective for conservation is needed; consequently it is important to systematically assess existing conservation

practices and to analyze possible opportunities of interaction and complementarities of systems components.

5. *How is PGRFA to be measured and assessed?* Genetic diversity is defined in quite different ways. Consequently, the estimation and valuation of diversity depends on the definition used. A task of economic valuation of PGRFA conservation is the analysis of the benefits arising from conservation, which also facilitates a judgement about which level of investment in conservation now and in the future would be appropriate.
6. *Who are the direct and indirect beneficiaries of PGRFA conservation?* In the political discussion, it has been implied that the private breeding sector is the predominant beneficiary of PGRFA conservation as far as "benefit sharing" is concerned. This has to be analyzed in more detail; the spread of benefits across society may be much broader.
7. *What systems of incentives may conduce PGRFA conservation effectively and what are their limitations?* PGRFA will not be conserved without appropriate incentives for those involved in conservation. Because of the characteristics of PGRFA, incentive mechanisms have to be created and analyzed in regard to their different levels of costeffectiveness.
8. *Which are the institutional mechanisms for PGRFA conservation at national and international levels and what are the constraints to their effectiveness?* Since the Convention on Biological Diversity entered into force in December 1993, the previous institutional system for the conservation and utilization of PGRFA has to be adjusted - several institutional mechanisms are outmoded. Issues of regulatory policy have to be analyzed, mainly the scope and limitations of markets cum / versus other institutional control mechanisms conducive to a conservation and sustainable utilization of PGRFA
9. *What are the main cost components for PGRFA conservation and what scope of cost-cutting measures enables an improved cost-efficiency?* For the analysis of efficient conservation systems, the cost elements have to be analyzed. The introduction of cost-cutting measures should improve the costefficiency of conservation activities.
10. *What are the costs of current global conservation activities and what may be "fair" burden sharing between the supply and demand side of genetic resources taking into account the distribution of benefits and the "capacity and willingness to pay"?* It is important to estimate the costs of current global and national conservation activities. Furthermore it has to be explored which conservation activities underlie whose financial responsibility.

This study wants to contribute to the development and the conceptualizing of more efficient conservation strategies by discussing the different aspects of PGRFA conservation and focusing on the cost aspect.

1.3.2
Approach and Data Sources

The discussion of these research questions will be divided into two steps:

In a first step, it is necessary to identify the PGRFA conservation objectives, the service needs for the genetic resources utilization and the institutional framework of conservation and transaction of genetic resources. A review of the literature as well as cooperation in the preparation process for the "International Technical Conference on Plant Genetic Resources" in Leipzig, 1996, organized by FAO, were the main sources of the general and specific information needed. A major source of information used here are the Country Reports, submitted by 154 countries in the preparation process for the conference.

The second step is to identify and collect data on the costs of the different conservation strategies and the institutional framework. In view of the inherent difficulties and limitations in compiling cost data on PGRFA conservation, the purpose and scope of the data collection were to obtain orders of magnitude of the current efforts made by the involved organizations and various other funding sources:

- Expenditures of funding and executing agencies active in conservation and utilization of PGRFA were calculated by gathering information from personal communications and documents. The information received represents most of international agencies; information was, however, not available for some international agencies, predominantly non-governmental organizations (NGOs).
- As of May 1996, 43 countries' focal points from 154 countries surveyed had replied to a questionnaire requesting financial information about national conservation activities. Among those responding were countries thought to have substantial programs in PGRFA (inter alia, the USA, France, Germany, Russian Federation, UK, Japan, China, India, Brazil, and Ethiopia), as well as a number of countries with smaller programs. Because of lacking information from governments and international institutions, expenditures for 1995 could not be calculated precisely. The national and international data were estimated based on the available information in order to obtain an order-of-magnitude estimate of total expenditures.
- Personal visits to and cooperation with selected genebanks enabled the calculation of the costs of genebanks in various environments.
- The costs of conservation activities were extracted from relevant case studies in the literature or were calculated based on expert judgements on its components.

1.4 Study Overview

Chapter 2 analyzes the state and dynamics of biodiversity generally and those of agrobiodiversity specifically. The development and extent of genetic losses are discussed, whereby critical analysis of the literature and methodologies of recording of genetic losses is necessary. In addition, the causes and determinants of the loss of plant genetic resources are analyzed. The main characteristics of specific conservation objectives, conservation methods, and economic valuation show significant differences between biodiversity and agrobiodiversity as elaborated in this chapter. Different methods of conservation are summarized to give the necessary technical background.

The conceptual framework for this study is described in Chapter 3. After a general overview, the concept for economic valuation of natural resources are presented in general and of PGRFA in particular, highlighting the benefits of conservation. As will be shown in Chapter 3, the loss of genetic resources is much determined by missing incentives for the conservation and transaction of PGRFA.

The institutional economic aspects of the transaction of PGRFA between supply and demand are discussed in Chapter 4. The analysis concentrates on the actors in the conservation and transaction activities. Furthermore, the role of the intellectual property rights in the exchange system are analyzed and the different institutional systems of exchange are discussed. In Chapter 5 the costs of plant genetic resources conservation are estimated. This is done on a global and a national level. Two case studies are reflected separately in Chapter 6 - the Indian and German national program for the conservation of PGRFA. Economic considerations and conclusions for PGRFA conservation policy are discussed in Chapter 7.

2 Genetic Resources: Status, Development, Losses, and Conservation Management

This chapter highlights the state and development of the world's genetic diversity centers, differentiated according to biodiversity in general and agrobiodiversity in particular. The extent and determinants of genetic extinction will be analyzed, and the differentiation between biodiversity and agrobiodiversity is discussed. In the last section, the methods for PGRFA conservation will be introduced.

2.1 State and Development of the World's Genetic Diversity Centers

There is a wide variation in the estimation of the overall amount of existing species or organisms: depending on different methodologies, the estimations lie between 5 million species (Stork, 1993) and more than 360 million species (André et al., 1994). This variation is determined by the limited knowledge about animals; especially as regards insects, the richest and most diverse group (Wilson, 1992), whereas the numbers of mammals and birds are fairly well known (Reid et al., 1993). The knowledge of the diversity of vascular plants is, however, relatively comprehensive, even though plants have not been inventoried as well as mammals or birds (Groombridge, 1992).

The inventoried plant species are not evenly distributed over the earth. According to OECD, (1996b) biodiversity - characterized as species richness - tends to be higher in: (1) warmer regions than in colder ones; (2) wetter regions than in drier zones; (3) less seasonal areas than in seasonal areas; and (4) areas with more varied topography and climate than more uniformed areas.

The differences in the distribution of plant biodiversity emphasize the importance of tropical and subtropical regions (especially tropical forests), and underline the significance of developing countries in terms of conservation and supply of plant genetic resources. Table 2.1 shows that regions dominated by developing countries, i.e., South America, Africa, and Asia, host high numbers of plants, whereas industrialized countries, e.g, European and North American countries, contain less than 5% of all documented plant species. South America, where nearly 30% of all higher plant species are to be found, is the most biodiverse region for higher plants, if biodiversity is defined simply by species richness.

Table 2.1. Distribution of higher plants by region

Region	Number of species[a]	Percentage of plants in a region to the overall amount of 300,000 documented plants
South America	85,000	28%
Asia	50,000	17%
Africa	45,000	15%
North America	17,000	6%
Australia	15,000	5%
Europe	12,500	4%

[a]: The number of species existing in the different regions has mostly been estimated provisionally, because of the lack of reliable surveys. This explains that the total of the given species' numbers does not even add up to the approximately 300,000 documented species, although the total should to be higher than 300,000 because of the double counting of species which are existent in two or more regions.
Source: compiled after data from Groombridge, 1992

In addition to the different regional distributions of species, significant plant diversity can be located in specific centers, which are differentiated according to the plant's affiliation to the group of plant genetic resources which are relevant to food and agriculture or to the group of not yet utilized plants. The regions with the main centers of general plant biodiversity and of diversity of domesticated plants (the so-called Vavilov's centers) will be characterized in the following section.

2.1.1 Centers for Plant Genetic Diversity

In order to locate areas of diversity, it is essential to define the criteria, which designate the quality of plant biodiversity in a given area. There are different approaches to present the state of biodiversity according to the objectives or the methodology chosen. Following the centers concept outlined in Davis et al. (1994), the criteria needed to define a center of plant diversity are (1) the biodiversity index, including species numbers and some biological population parameters; (2) the number of endemic species restricted to a given area; (3) the genetic separation between the species of a given area; (4) the grade of taxonomic scattering; and finally, (5) the current or potential economic value of the species. At present, all 5 criteria may be utilized in selective areas, where all information exists or can be obtained within a reasonable time frame. To define the centers of plant biodiversity on a global level, the operational use of the criteria is decreasing from the first to the fifth criterion.

After evaluating 1,400 floras, floristic studies, bio-geographical essays and vegetation studies, Barthlott et al. (1996) mapped out the species richness in ten "diversity zones", giving the numbers of species per 10,000 km^2. In comparison to a more simple map of Malyshev (1975), the map of Barthlott et al. is the first map outlining the zones of biodiversity on a global level. Six global diversity centers with more than 5,000 species per 10,000 km^2 were identified; all were situated in the humid tropics and subtropics: Chocó-Costa Rica Center, Tropical Eastern Andes Center, Atlantic Brazil Center, Eastern Himalaya-Yunnan Center, Northern Borneo Center, New Guinea Center.

Although there was an increase in the number of flowering plant species over the past 140 million years, the changes in the last 20,000 years were slight; the extent of extinction of species was more significant than the amount of emerging of new species (Signor, 1990).

2.1.2 Centers of Agrobiodiversity

The Russian botanist and geneticist N. I. Vavilov was the first to notice that diversity within agricultural crops was not equally dispersed (Vavilov, 1926). Vavilov defined the regions with the highest diversity within one crop as the centers of diversity of crop plants. He assumed them to be the centers of origin of that crop (now called "Vavilov's centers of origin"). He mapped out 8 major centers: East Asia, Tropical Asia, South-western Asia, the Near East, Mediterranean, Abyssinia, Andean, and Central America including South Mexico. The continued geneflow between crops and their wild relatives in these areas underpins their importance as sources of new variability. Different species of the same crop, however, have been domesticated in different places[7], or the independent domestication of the same species of the crop occurred in various places[8] (Harlan, 1976). Consequently, one single center of origin cannot be identified for all relevant crops. Additionally, because of the evolutionary process combined with human activities (Juma, 1989), secondary centers of diversity were created (see Table 2.2), with significant variations to those from the centers of origin (Harlan, 1971).

Zeven and Zhukovsky (1975) enlarged Vavilov's 8 centers of origin and diversity by Harlan's secondary centers of diversity (Harlan, 1971) and drew the mega-centers of cultivated plants to describe the origin and the main region of the present diversity of the major crops on global level.

Although tropical forests are the areas singled out for rich biodiversity, the drier ecosystems are far more important for crop resources, which are situated in semi-arid and mountainous areas and are agriculturally marginal with respect to soil fertility, water, and so forth (e.g., the Near East and Central Asia).

---

[7] Different species of yams were domesticated in West Africa, Southeast Asia, and in tropical America (Harlan, 1976).

[8] Both cassava and sweet potato were domesticated independently in Central and South America (Harlan, 1976).

Table 2.2. Examples of important secondary centers of diversity

Crop	Primary center of diversity	Secondary center of diversity
Finger millet[a]	East Africa	South Asia
Sorghums[b]	Africa	South Asia and Latin America
Beans and groundnut	Latin America	Sub-Saharan Africa
Maize and cassava[c]	Latin America	Indochinese-Indonesian area
Barley	Fertile Crescent of the Middle East	Ethiopia
Rye	Near East and Mediterranean	Introduced as weeds in northern European fields
Oats	Near East and Mediterranean	Introduced as weeds in northern European fields
Rubber[d]	Latin America	Indochinese-Indonesian area

Source: FAO, 1996a; [a]: Engels et al., 1991; [b]: Kloppenburg and Kleinman, 1987; [c]: Plucknett et al., 1987; [d] Wood, 1988a

Although the biodiversity declined over time because of human interaction, the level of agrobiodiversity increased at the same time due to human activities. Agricultural domestication, breeding, and trading of crop species increased the genetic diversity as well as the diversity of varieties and - to a certain extent – the diversity of crop species over the past 10,000 years (McKelvy Bird, 1996). While the number of crop species utilized in agriculture is relatively small, the diversity within such species is often immense. For instance, there are an estimated 130,000 distinct varieties within the rice species *Oryza sativa* (Chang, 1992). With modern, scientific breeding emerging, the expansion process changed its trend: the main objective for exchange of varieties changed from adapting an imported crop or variety to local conditions to incorporating only some traits of a foreign variety into existing, local or modern varieties. These modern-bred varieties include genetic resources from completely different sites and are spread over a larger area than were the former local varieties, thus reducing the local intra-crop diversity and the inter-crop diversity.

2.1.3 The Interdependence of Countries on PGRFA

The centers of origin often differ, however, from today's important areas of production. Secondary centers of diversity were established in some areas. Consequently, over two thirds of developing countries acquire more than half of their crop production from crops domesticated in other regions (Wood, 1988a). The impact of species originating from other regions on local food production for major crops can be seen in Fig. 2.1. Only South Asia, South East Asia, and West Central Asia depend on food crops originated from other countries for less than 50% of their food production. In all other regions over 50% of the food production is based on species originated from other regions (Kloppenburg and Kleinman, 1987). Regions like Australia, North America, the Mediterranean, and Europe depend on crop species from other regions for over 90% of their food production; but also Africa shows over 87% dependency. The interdependence of crop species between regions and countries is demonstrated by crops such as cassava, maize, groundnut and beans. These crops originated in Latin America, but have become staple food crops in many countries in Sub-Saharan Africa (Plucknett, 1987). On the other hand, African-originated millets and sorghums make a considerable contribution to other areas such as South Asia and Latin America (Kloppenburg and Kleinman, 1987).

Fig. 2.1. Percentages of food production of major crops based on species originating from other regions

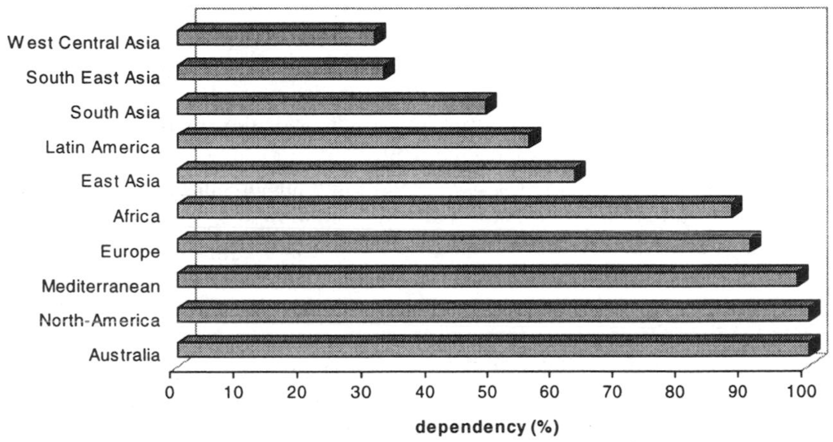

Source: adapted from data according to Kloppenburg and Kleinman, 1987

2.2 Extent and Determinants of Genetic Extinction

The diversity of plants is said to be declining. This statement forms the basis for various efforts on national and international level to conserve the existing diversity. But in order for efficient activities to maintain an optimum of diversity, it is necessary to analyze the decline of biodiversity. Is it a decline on all different levels of biodiversity? Is it mainly a reduction in intra- or inter-genetic diversity? In the following section, the character and extent of biodiversity reduction for plants and especially for agriculturally relevant plants will be analyzed and the determinants of the decrease in biodiversity will be discussed.

2.2.1 Review and Critical Analysis of the Methodologies of Recording of Losses

Because of the significant differences between plant biodiversity in general and agrobiodiversity (see Chap. 2.3), the decline in general plant biodiversity and its determinants will be discussed very briefly, before the reduction of agrobiodiversity and its underlying causes are analyzed in more detail.

2.2.1.1 Genetic Reduction in Biodiversity and its Determinants

The physical decline and extinction of plant species, which causes genetic losses as well as the transformation of plant species over evolutionary time, is a natural process. On the one hand, the history of extinction in vascular plants shows some major catastrophes, which had a significant influence on the structure and composition of the vegetation and the extinction of species (Knoll, 1984). On the other hand, the process of plant extinction is characterized by competitive displacement by plants more able to adapt and the gradual adaptation of environmental changes (Groombridge, 1992).

The primary causes for extinction of the plant species in recent history are the human-induced habitat modification or fragmentation and habitat destruction (see Fig. 2.2). Furthermore, other human activities may have either a positive or negative impact on the existing biodiversity. The import of some new species into a habitat may enrich biodiversity, if the exotic species finds a niche, or it may reduce biodiversity if it has comparative advantages over traditional species.

In addition to these deterministic processes, stochastic processes affect the population dynamics of species as well. According to Schaffer (1987), stochastic processes are determined by (1) demographic uncertainty as a result of random events in the reproduction and survival of individuals; (2) genetic uncertainty as random changes in genetic make-up; (3) environmental uncertainty caused by unpredictable changes in weather, pests, and nutrition supply; and (4) natural catastrophes, e.g., fires, floods, and droughts. The impact of these stochastic processes on the level of biodiversity in a specific habitat does not always imply a decline in biodiversity. The four underlying factors may as well maintain or even improve the biodiversity. For instance, forest fires improve the regenerative ability of the sequoia trees in California or enable the germination of certain plant species which have been suppressed by other dominating plant species.

In addition to the direct impact of human intervention, the interaction of human intervention with the stochastic processes is causing an over-proportional loss of species. This process of isolation and fragmentation of populations leads to an accelerated and over-proportional extinction of species (Schaffer, 1987).

Difficulties in recording extinction rates of plants are caused by the lack of documentation of existing species at a given time in the past. It is difficult to unequivocally demonstrate that extinction has occurred, because many species may continue to persist for years in the seeds before germinating (Jenkins, 1992).

According to the Convention on Trade in Endangered Species of Wild Fauna and Flora (CITES), only those species may be cited as extinct which have not been recorded for at least 50 years (CITES, 1993). Consequently, current extinction is seldom documented yet.

Fig. 2.2. Impact of various activities on present biodiversity

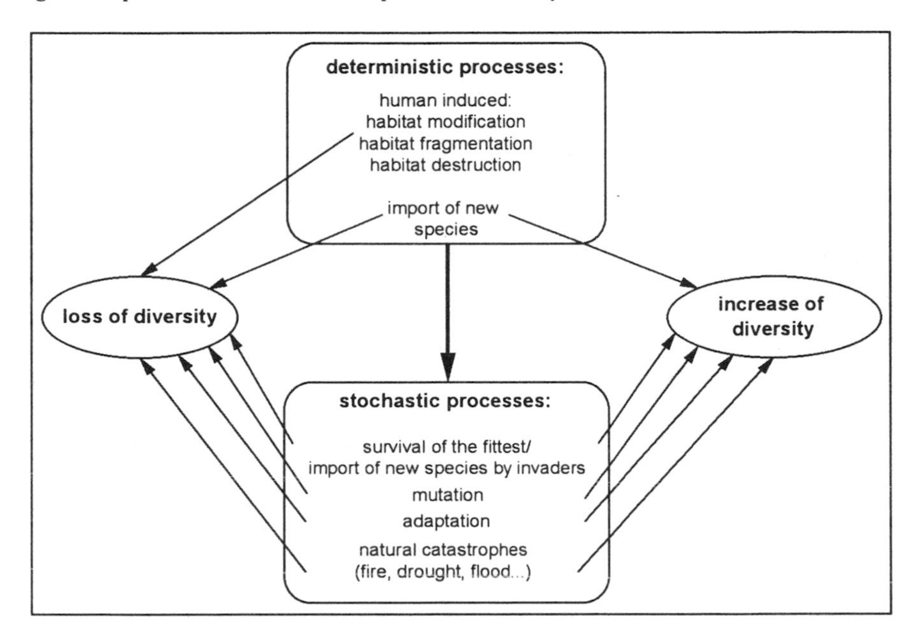

Because of the difficulties described, estimates for present and future extinction rates for plants are not based on reported and documented extinction, but rather on models. These incorporate extrapolations from estimated habitat destruction, combined with bio-geographic assumptions derived by relating the numbers of species to the area of habitat. Another major difficulty in recording the decline of biodiversity is the determination of a causal link between human activities and extinction of species, which is necessary for the estimates. (Jenkins, 1992).

For the most part, the studies have been based on the estimated species richness of tropical forests combined with their known and estimated deforestation rates. Estimates of the extinction rate of plant species range between 1% and 11% per decade depending on the assumed rate of deforestation and the relation between the number of species and the existing area (see Table 2.3). The different results depend to a large extent on the prospected development of the available area, which determines the relation between the number of species and the existing area.

Some major shortcomings in this approach must be considered when citing the estimated values. These estimates include plant as well as animal species. Hence, it must be questioned whether the same relationship can be utilized for plants as for animals. Because of different population dynamics and area requirements, it seems that the predicted overall estimates could be reduced even further for estimated plant extinction rates (Groombridge, 1992). Although the majority of terrestrial species are derived from the tropical forests, it seems insufficient to calculate the overall extinction rate of plant species based only upon the further development of the tropical rain forests.

Jenkins (1992) points out that species richness as well as habitat destruction in the forests is not distributed evenly. As shown by Dobson et al., the fact of the "hot spots" of potential extinction must be reflected in these calculations. After examining patterns of geographical distribution of 924 threatened and endangered plant and animal species in 2,858 U.S. counties, Dobson et al. came to the conclusion that most endangered species in the U.S.A. are only found in relatively few critical geographical areas - the "hot spots" of potential extinction (1997).

Furthermore, the estimates assume that forest conversion means total habitat destruction. Taking the different conversion possibilities mentioned above into account, the habitats are, however, often modified without resulting in the total extinction of given species. In spite of these open questions, there is a constant or even accelerating destruction of tropical forests and other ecological marginal areas, which has a negative impact on the diversity of plant species in these areas, regardless of the impossibility of even precisely quantifying the extinction rate until now.

Table 2.3. Estimated current and future extinction rates of plant species

Global loss per decade	Estimates	Method of estimation	Reference
8 – 11%	Between 1980 and 2000 loss of 15-20% of species	Estimated species-area curve, forest loss based on global 2000 projections	Lovejoy, 1980
8%	By 2015 loss of 2000 plant species per year in tropics and subtropics	Estimation based on deforested area by 2015	Raven, 1987
7%	Between 1990 and 2000 loss of at least 7% of plant species	Estimation based on species loss in 10 hot spots	Myers, 1988
2 – 5%	By 2020 loss of 5 – 15% forest species	Forest loss based on twice rate projected by FAO for 1980-85	Reid and Miller, 1989
1 – 5%	Between 1990 and 2015 loss of 2 - 8%	Forest loss based on current rate and 50% increase	Reid, 1992

Source: Reid, 1992, Ehrlich and Wilson, 1991

2.2.1.2
Genetic Extinction in Agrobiodiversity

It is essential to define the underlying unit for PGRFA for the assessment of the development of agrobiodiversity over time, as well as for an economic discussion on the benefits and costs of PGRFA conservation. Diversity is characterized by the genetic resources existent in a given framework as shown in Chapter 1. The terminology of "plant genetic resources for food and agriculture" does not, however, permit applications of economic concepts to the problems of scarcity, conservation and transaction. Agrobiodiversity is often equated with richness in crop varieties. In fact, crop varieties are the principal indicator for agrobiodiversity-richness as well as the economic unit of benefit valuation. But the correlation between phenotypic (expressed as differences in varieties) and genetic diversity (on gene level) does not seem to be positive. After analyzing the genetic diversity in some crops, and in spite of demonstrating much phenotypic diversity, Clegg et al. (1992) found out that these crops had relatively narrow primary and cultivated gene pools due to severe genetic bottlenecking, either during the domestication process or later in the crop's evolutionary history. As Smale states: *"Plant populations that appear different may in fact carry the same genes, and populations that appear the same may carry different genes."* (1997, p. 1259); this may lead to one conclusion: the same amount of varieties in two different crops does not necessarily mean that the diversity of the two crops is equally high in the region. The total genetic variation of all the different varieties of one crop may be much less than that of the second crop, although the second crop may have less varieties.

The conservators of PGRFA (the major actors of the supply side of PGRFA) tend to favor the use of qualitative traits as marker genes to monitor the extent of diversity. Often not agronomically relevant but frequently genetically linked to agronomic traits, these traits are mostly components of the characterization data, giving information on color, morphology or enzyme variants of accessions. The major actors on the demand side of PGRFA (the breeders and the biotechnology industry in general) are, however, more interested in quantitative traits, including agronomic traits such as yield capacity or plant height. These quantitative traits, which define the breeding goals, are lastly functions of certain biological organisms. Often these traits are not due to single genes but rather to a combination of genes representing one required function. For this reason Vogel (1994) argues in favor of using the term "genetically coded function" (GCF) as a basis for economic valuation of genetic resources exchange mechanisms. Although this seems correct, GCF are determined by specific combinations of genetically coded information (GCI). Finally, the actors (especially the biotechnology industry), demand information, which determines certain functions. As technologies improve, one objective will be the virtual construction and reproduction of genetic basis (e.g., production of aminoacid sequences). Consequently, in light of the emerging market, and the articulation of the demand side, genetically coded information is recommended here as the unit, which can be utilized for economic and - in the long run - institutional analysis, discussion and negotiations of PGRFA. Genetically coded information will be the unit which will enable the development of appropriate concepts for identifying alternative institutional mechanisms for the protection, transaction, and utilization of genetic resources within the framework of general concepts of nature conservation, including the formation of markets.

Another advantage of genetically coded information as an economic unit is the opportunity to have a comprehensive unit which is applicable to genetic resources exchange and transaction systems in agriculture as well as to the transaction of genetic resources in general. As a future consequence, the institutional framework of conservation of genetic resources will be challenged by an emerging information market. The major disadvantage of the newly introduced economic unit of genetic resources is that - for the time being - it is not practical; this is especially true for PGRFA, because the technology needed to identify genetically coded information is not advanced enough yet for general application. Consequently, taking into account that varietal diversity is not the best unit, but the most convenient, the varietal diversity is the most practical indicator for PGRFA diversity as well as for the benefit and cost assessment for the present analysis.

A narrowing of the species diversity in agrobiodiversity on global level can be stated following the given definitions of agrobiodiversity (see Fig. 2.3.): Among an estimated 400,000 plant species (Groombridge, 1992), approximately 300,000 species have been documented (Heywood, 1995). Ten % (i.e., 30,000) of these plants are edible and about 7,000 have been cultivated or collected by humans for food or other agricultural purposes at one time or another (Wilson, 1992). The production statistics of the international organizations mention approximately 200 crop species. Only 30 of all these mentioned crops, "feed the world", i.e., they provide together 95% of dietary energy or protein. Only three crops, however, rice (26%), wheat (23%), and maize (7%), provide over 50% of the global food energy supply derived from crops (FAO, 1996a).

The narrowing of the species diversity is not only evident in crop production and in farmers' fields, but also in the investment in conservation activities (see Appendix 1 and Appendix 2). Approximately 25% of all 6.2 million ex situ stored accessions come from one of these three crops (WIEWS, 1996; FAO, 1996a). Furthermore, at least 37% of all stored accessions in the CGIAR centers are derived from one of these three crops. Additionally, the CGIAR system spent 25% of all its 1995 expenditures for the genetic resources activities of the 9 relevant centers for activities relating to wheat, rice and maize (SGRP, 1996).

In addition to the narrowing of the species diversity, the varietal diversity in agrobiodiversity seems to be decreasing. The loss of genetic diversity is a major incentive for conservation efforts; there have been, however, only a few systematic studies providing quantifiable estimates of the actual rates of reduction in agrobiodiversity. There is neither information on historical situation as regards agrobiodiversity, nor a comprehensive inventory of what currently exists in farmers' fields. As an indicator for the loss of varietal diversity, cultivar replacement measured by area is quoted in some of the Country Reports as well as in some other literature (see Appendix 3).

Fig. 2.3. The narrowing of species diversity in agrobiodiversity

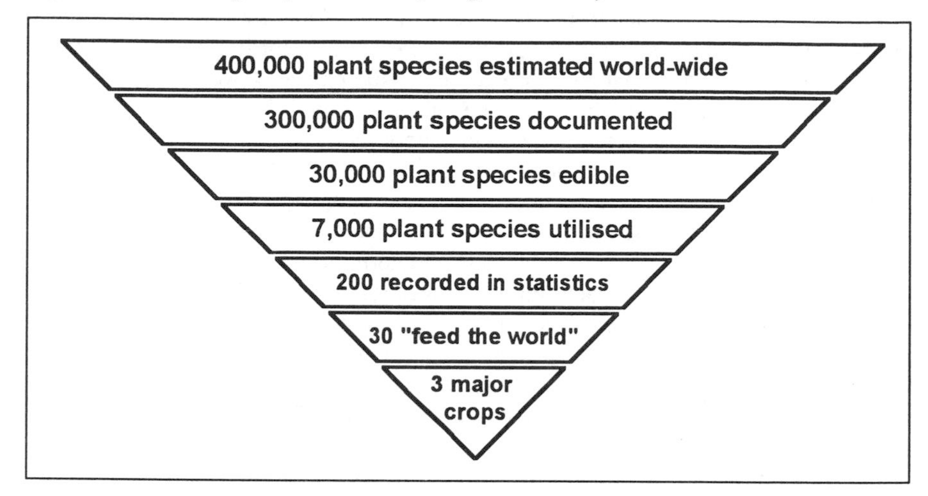

Source: according to data from Heywood, 1995, Wilson, 1992, FAO, 1996a

Nevertheless, taking into consideration that 154 Country Reportshave been analyzed, genetic extinction in agriculture is poorly documented. The use of cultivar replacement as an indicator in estimating varietal loss is not precise, because studies indicate that because of private insurance considerations or other expected benefits some farmers may continue to use traditional varieties even after adopting modern varieties (Brush, 1994). In spite of some examples recording the complete extinction of local varieties and the reduction of cultivated area for local varieties, it seems that the process of modernization and varietal replacement in most countries has not yet reached the stage like that recorded in the USA (see Appendix 3; Fowler, 1994). Until now, genetic extinction occurs mainly in the form of the replacement of traditional varieties in main production areas, where genetic uniformity is increasing instead. Because the process of concentration of diversity occurs in smaller areas than in the past the last resort for the majority of traditional varieties are the ecological marginal areas.

For example, 75% of an area which once accommodated up to 30,000 rice varieties in India is now taken up by only 10 (FAO, 1993a). The significant change in India's varietal diversity was, for example, not the amount of local rice varieties completely replaced by modern varieties, but rather the area under which the majority of local varieties is cultivated. In addition to the narrowing of the species diversity and the declining spatial shift in varietal diversity, the changes in the genetic diversity must also be analyzed as well. This is the most complicated task, especially because the functioning of single genes as well as gene combinations is not yet fully understood. Furthermore, a single cell of a flowering plant may consist of over 400,000 genes (Hinegardner, 1976).

Molecular analysis can be used to detect the changes in genetic diversity in a variety. It seems, however, that no clear consensus about the changes of the genetic diversity can be reached, as in bread wheat for instance (Smale, 1997).

Each new breeding activity has the principle of narrowing the genetic base of the source variety. On the other hand, incorporating new traits by introducing genes broadens again the variety's genetic base. For instance, according to Tarp (1995), today's genetic base of tomatoes is broader than 40 years ago. Another method in assessing the genetic variation in one given crop is to monitor genetic vulnerability. Genetic vulnerability is the genetic constitution of a crop, if it is uniformly susceptible to a pest, pathogen or environmental hazard (NRC, 1972). Genetic vulnerability is a consequence of decreasing genetic variation in a given area, mainly through the widespread replacement of diverse varieties by homogeneous modern varieties. Two factors determine the vulnerability: (1) the relative areas devoted to each cultivar, and (2) the degree of uniformity (relatedness) between cultivars. The uniformity in modern varieties is higher than in landraces. Genetic vulnerability creates a potential for widespread crop losses because of the increased susceptibility to pest, pathogen or environmental hazards. The 1840s' pandemic of late blight in potatoes (*Phytophtora infestans*) in Ireland is a famous example of the danger of genetic uniformity. More examples of genetic vulnerability in countries are given in Table 2.4. Genetic uniformity of crops will increase in the future through the improvement and increased application of biotechnology in agriculture.

The possibility of losing genetically coded information from farmers' fields is increasing because of the threat of losing traditional varieties and crop species. Therefore, the irreversible loss of genes, as well as the loss of gene complexes, plant varieties, and crop species are feared.

Consequently, this threat is the major incentive for the increased conservation activities of the last 30 years. As has been discussed in this section, a narrowing in species diversity utilized for agriculture and a decline in crop varietal diversity can be pointed out. Even though a significant drop in the landraces' share of all agricultural varieties can be seen since the turn of the century (Fowler and Mooney, 1990), the concentration of area among the top cultivars is lower at present than it was at the summit of the Green Revolution period (Smale, 1996b).

Agrobiodiversity is not decreasing overall, but a decline of specific diversity at specific places at a specific time can be seen - i.e.:

- varieties were always used and never conserved by farmers for the longterm[9];
- the utilization of agrobiodiversity - and thereby its steady change - was the only method of conservation;
- the intra-species diversity may be reduced according to the number of different varieties grown on farmers' fields due to the replacement of traditional varieties by modern varieties; not all the genetic information of these old varieties is lost, however.

---

9 Farmers usually carry out ex situ conservation only by storing seeds from one harvest to the next season.

Table 2.4. Reported genetic vulnerability in agricultural crops

Country	Reported genetic vulnerability
Bangladesh[a]	Over 67% of the wheat fields were planted to a single wheat cultivar "Sonalika" in 1983, when wheat as a crop had expanded from a very low level in the rotation
China[b]	Genetic uniform F1 hybrids of rice covered 5 million hectares in 1979 and covers 1990 15 million hectares
Europe[c]	In European barley protection against mildew is increasingly dependent on one gene and one fungicide
former Soviet Union[d]	in 1972, the winter wheat cultivar "Bezostaya" was grown over 15 million hectares
India[e]	30% of the wheat fields were planted to a single wheat cultivar "Sonalika" in 1984
Indonesia[f]	74% of rice varieties descended from one maternal parent
Netherlands[g]	One barley variety/cultivar accounts for 94% of the spring barley planted
Sri Lanka[f]	75% of rice varieties descended from one maternal parent
USA[h]	75% of potato in 4 varieties
USA[h]	50% of cotton in 3 varieties

Source: Country Reports of China;
[a] Dalryample, 1986; [b] National Research Council, 1993; [c] Wolfe, 1992; [d] Fischbeck, 1981, [e] Plucknett et al., 1987; [f] Cabanilla et al., 1993; [g] Vellve R 1992; [h] NRC, 1972

A loss of single genes and of the unique combination of genes, however, could not be proved. Based on the present scientific knowledge, it is impossible to ascertain whether the genetically coded information of agricultural crops has decreased, maintained its level, or even increased over time. The link between genetically coded information and the utilized varieties has not yet been understood. As a result, important questions for the conservation activities remain unanswered:

- How high is the rate of unique information in one variety?
- How much unique genetic information is lost by losing varieties in the field?
- Is there a link between varieties and genetically coded information at all?

Despite the above-mentioned uncertainties, the conservation of PGRFA is taking place, determined by the conservators worldwide and the declarations of intent by politicians at various fora, which were explicitly documented in the "Leipzig Declaration" as an outcome of the International Technical Conference on Plant Genetic Resources, held in Leipzig, Germany, in June, 1996 (FAO, 1996b).

2.2.2 Determinants for the Loss of PGRFA

The level of PGRFA diversity maintained in farmers' fields is determined by the decisions made on the farm level, depending on the individual or farm-specific objectives. The practice of cultivating different varieties or crop species and thereby maintaining a specific level of PGRFA diversity, is mostly a positive externality of the farm sector. The decisions made at the farm level concern the degree of change of the existing agricultural system on the farm level. The changes in the agricultural system determines the level of land-use development, which has the most severe impact on the future level of PGRFA diversity in farmers' fields. Furthermore, the maintenance-level of PGRFA in farmers' fields is not determined by the land-use development and farmers' decisions only, but also by related factors; among others, the socio-economic factors, the development of a relevant market, as well as policies and institutions. These factors influence the decision making process at the farm level and are summarized as the factors at the framework level, which directly or indirectly influence the present and future level of PGRFA in farmers' fields. Consequently, the decision-making process on the farm level is the key element of the factors determining the diversity level of PGRFA in farmers' fields. depicts the three levels: the framework level, the decision-making process on farm- evel, and the level of the land-use development, whereby the main factors on the framework level are specified. The site- and time-specific factors of the different levels and their inter- and intra-linkages determine whether the quantity of PGRFA diversity in farmers' fields may be maintained, increased or reduced. In the following section, the different factors of the different levels shown in are discussed in more detail.

The farmers, being the main contributors to diversity of PGRFA in the past, are at present those who influence the state of diversity of PGRFA the most with their day-to-day activities. The decisions at the farm level may result in changes of the agricultural production system. These changes will influence the *land-use development.* The major change in the agricultural system is its intensification through the application of new technology, which may lead to variety replacement, over-exploitation of genetic resources, and habitat destruction (see Fig. 2.4). As a side-effect, the changes in the agricultural systems may lead to the introduction of new pests, diseases or weeds; or, in an extreme case, these changes may lead to the abandonment of agricultural production and the development of non-agricultural land-use systems.

All the different kinds of land-use conversion described above are determined mainly by the decision-making process at the farm level. These decisions are influenced by the farmer's constraints and possibilities of land-use activities on the one hand, and by economic, political, and social factors on the other. This broadens the concept and integrates factors at the *framework level*, which influence the extent of PGRFA in farmers' fields. The factors are summarized in the following section (see Fig. 2.4).

Fig. 2.4. Factors determining the diversity-level of PGRFA in farmers' fields

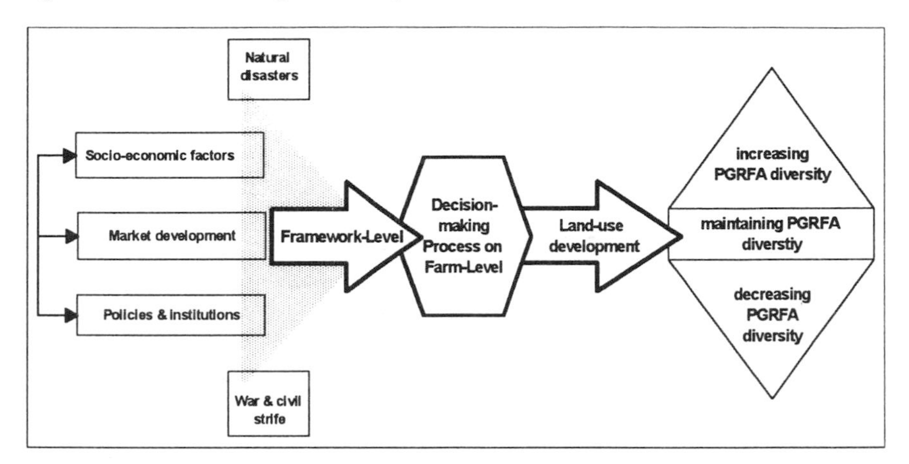

Many agricultural areas - are not threatened by variety and species replacement by the farmers so much, but rather more by natural disasters, in addition to the over-exploitation mentioned above. Natural disasters. e.g., droughts or floods, may destroy the whole crop population, resulting in the loss of specific genetic resources. Such effects have been reported by Cameroon, Burkina Faso, Guinea, and Kenya, for instance (respective Country Reports FAO, 1996h). In addition to natural disasters, war and civil strife, landraces can be wiped out by neglecting the field work, by consuming the last seed, and by moving to safer areas and leaving the fields and seeds behind. In countries like Rwanda, Somalia, Angola and Cambodia civil strife and war have contributed significantly to genetic extinction. Genetic extinction seems to be one of the consequences of the large-scale use of defoliants during the war in Vietnam (respective Country Reports).

The main socio-economic factors determining the level of PGRFA maintenance are population increase and population density, the purchasing power of the consumers, poverty of the rural population involved in agriculture, the available infrastructure, and the development of technology. The general consumption pattern is determined by the development of the population as well as by the development of the purchasing power. The demand for food generally increases with population growth, whereas the increased purchasing power is reflected in a higher quality of food. Consequently, an increasing demand for more quantity and quality of food leads to an increased demand for higher agricultural production and productivity, which results in the already mentioned increasing homogenization in agricultural production (Swanson, 1996).

In developing countries, the poverty of the rural population involved in agriculture generally induces the maintenance of traditional varieties, because they are too poor to invest in modern technologies, i.e., seeds, fertilizer, pesticides and general production systems. Furthermore, the poor are often already situated in or are pushed into marginal areas, where modern varieties are not competitive.

Fig. 2.5. Land-use development leading to the decline of PGRFA diversity

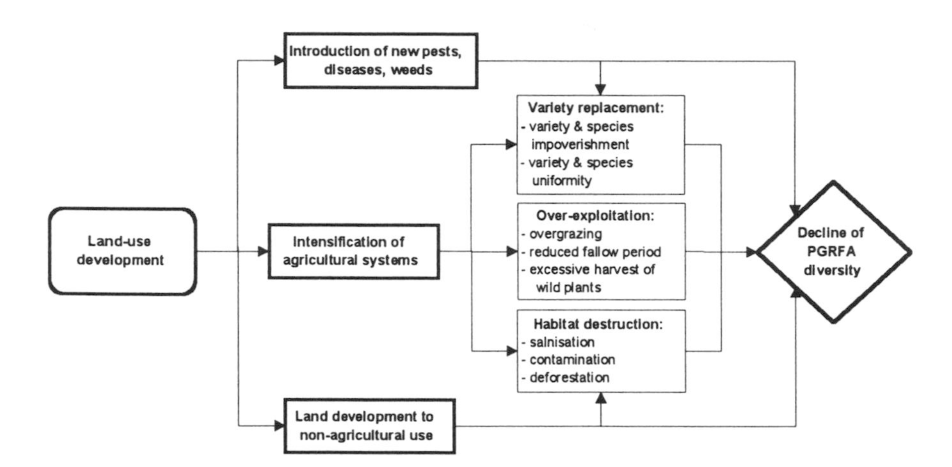

As Barbier et al. point out (1994), the poorest groups of developing countries often concentrate in the most ecologically fragile zones of their countries. Therefore, the resource-poor farmers normally maintain a higher level of diversity than do resource-rich farmers, generally located in higher potential areas. This general theory does not exclude inverse behavior by resource-poor farmers, as for instance described by Weltzien-Rattunde (1996) for Rajasthan and Gujarat in India. In the semi-arid to arid regions, resource-poor farmers buy hybrid millet seed for economic reasons. If these farmers have no seed left over from the last season, they will borrow seed of other crops from their neighboring resource-rich farmers. Because they cannot borrow all of the seed they need, they buy the less expensive hybrid millet seed, as the only millet seed for sale at all.

The situation of the farmers is closely related to the availability of infrastructure and the appropriate technology. Without the physical infrastructure, e.g., roads and more sophisticated communication systems, farmers do not have the access to markets; especially to markets for new technologies (e.g., seeds of modern varieties) and for information like extension and research. Without sufficient integration into the most important markets, the replacement rate of species and varieties will be very low or even will not exist at all.

Summarizing the socio-economic factors determining the level of maintenance of PGRFA in farmers' fields, the main factors influencing the replacement are those known from the diffusion discussion of technology adaptation in developing countries (e.g., Binswanger and Ruttan, 1978).

The development of a market for PGRFA is still in its infancy, because of PGRFA's characteristics of a free and public good as well as the legal uncertainty concerning the property rights. The demand for PGRFA has, however, existed since the beginnings of agriculture, and is increasing as a consequence of developments in conventional plant breeding and as well as in biotechnology (von Braun and Virchow, 1997). At the same time, the natural supply of PGRFA is vanishing because of the decline in the diversity of species and varieties. The decrease in supply is the driving force for the development of a new, potentially large market, in which a former free public good will be traded as a privately owned, scarce good. But as long as the potential traded products are not well-defined, the property rights and the exclusiveness cannot be guaranteed; consequently the market cannot function well yet.

In a situation where markets are emerging in a field full of complex problems and where people with very different levels of knowledge are involved, such insufficiency does not come unexpectedly (Cowen, 1988). Therefore, as Cansier points out in regard to all natural resources, there is not a "market failure" of all necessary allocation mechanisms for PGRFA, but rather a policy and institutional failure (Cansier, 1993). It is the failure of governments to set up institutional arrangements within which markets can deliver socially preferred outcomes. The lack of or the very slow development of those arrangements is reducing the momentum of the development of the PGRFA market.

Hence, policies and institutions represent the third major group of determinants having a significant impact on the extent and development of genetic resources diversity. Incomplete information and the uncertainty of the effects of PGRFA loss on global welfare as well as the costs of such loss handicap an economically efficient approach to diversity conservation. Rent seeking and the distortion of economic incentives are further points where policies determine the diversity of PGRFA. Policy support of modern varieties through governmental price control or subsidies for seed of modern varieties, promotion of fertilizer and pesticides will influence farmers' decisions[10]. Legislation to protect and guarantee the uniformity, distinction, and genetically stability of seed[11], i.e., certified seed, is at the same time hampering the trade and maintenance of old cultivars and farmers' varieties. Furthermore, the quality of political coordination, e.g., between countries of similar PGRFA diversity and with a similar availability of genetically coded information, determines the support of conservation efforts on national and international level.

As Swanson (1996) highlights, the changes in the insurance strategy determine the level of diversity as well. Because of changes in the insurance policies for farmers, the importance of diversity of PGRFA in the field has declined in favor of other insurance strategies. Today, farmers are taking advantage of monetary insurance, non-agricultural income, family relations, and national and international aid assistance. Crop insurance schemes are a well-established means of protection against crop losses in many developed countries, and, increasingly, in many developing countries as well. Governments provide public insurance systems which include cover against losses due to pests and diseases, as for instance India, the Philippines, Cyprus and Venezuela (Roberts and Dick, 1991).

---

[10] In Africa, the replacement of traditional agricultural practices by modern practices is common, as is policy support of high-yielding varieties to boost agricultural production (Information from Country Reports).

[11] See UPOV, 1992.

Other governments, e.g., the US government, subsidize private insurance companies (Hyde and Vercammen, 1997; Swanson et al., 1994).

The major institutional factor determining the maintenance of PGRFA diversity is the missing or poorly defined property rights for genetic resources. The public good characteristics which still exist give rise to free rider problems. Property rights are essential for the conservation of agrobiodiversity: the incentives to maintain or decrease the amount of PGRFA diversity depend on the structure of the various rights of all agents involved. Open access conditions prevail where no property rights exist.

The level of PGRFA diversity is determined by the existing framework, the present situation of markets for PGRFA, policies influencing the maintenance of PGRFA, and adequate institutions as well as the intensity and direction of land-use development. In addition to these two important factors, the *farm level* is the key position of the decision-making process influencing the diversity in situ.

If the farmer expects more benefits resulting from new varieties, which were improved on the basis of the old varieties, the farmer will replace the old varieties with the new varieties[12]. According to the field experience of Pundis (1996), farmers take over a new variety if the yield gains are over 15% more as compared to the traditional variety. The more the results from plant breeding are better (i.e., incorporation of all needed or asked for traits), the more farmers' varieties will be replaced by the improved varieties. The replacement may take place in one step; but as breeding is a long and difficult process, the replacement will be occurring over some time as well. For each trait incorporated in an improved variety, the advantage of one or more farmers' varieties will be diminished or lost.

Summarizing this section, it can be highlighted that most loss of biodiversity and agrobiodiversity is an incidental or unintended side-effect of other activities (McNeely, 1996). The extent of losses of PGRFA is less than expected and often described in the literature. The factors determining the level of agrobiodiversity may be found on three different levels, indicating a further decline in diversity unless a market for PGRFA can be developed, which incorporates an expected high value for PGRFA.

There is not only a loss of genetic resources in situ, but rather genetic resources are also lost in genebanks. Because of insufficient financial resources, mismanagement, and untrained staff the condition of stored genetic resources is far from being safe. For instance, at the genebank at Fort Collins, USA, 50% of all conserved accessions are not viable and only 28% of the accessions stored are still healthy and may be utilized in the future (Hobbelink, 1989).

---

[12] This generalized benefit optimizing on the farmer level is even not questioned by the findings of Bush and Meng, 1996 stating that farmers, even though introducing improved varieties are keeping farmers' varieties in their production systems. However, these farmers are doing this mainly because the introduced varieties do not have certain traits (i.e., taste) which are known in the traditional varieties. However, assuming that improved varieties may incorporate these specific traits as well, the farmers will have no benefits in keeping the traditional varieties any longer.

Another example comes from India, in 1995, when 5,311 original accessions of the Assam Rice collection collected in the 1960s had deteriorated or had been lost through unavoidable circumstances. The collection could be restored only because India had sent the duplicated material to IRRI (IPGRI, 1995).

2.3 Similarities and Differences Between Biodiversity and Agrobiodiversity

If biodiversity is used as a synonym for plant genetic resources in general and agrobiodiversity for PGRFA, it implies that agrobiodiversity is only a subsystem of biodiversity and that both can therefore be treated in one system. Although there are similarities between biodiversity and agrobiodiversity, the existing differences recommend a differentiation of both diversity forms. Both systems interact and overlap with each other, while each system has its own, characteristic features. For instance, modern agriculture is a major cause of the extinction of plant varieties for food and agriculture. Furthermore, modern agriculture causes the extinction of wild plant species (see Chaps. 2.2 and 3.2) because of monoculture, intensive use of agricultural chemicals, and over-mechanization. Nevertheless, the pressure on land has been reduced in industrialized countries only because of modern intensive agriculture and can be reduced in low-income countries in future as well. Therefore, causing the extinction of varieties of agricultural crops, modern agriculture provides area for the conservation of biodiversity by increasing the area productivity. In other words, intensive agriculture enables a reduction of the pressure to take non-agricultural land into production, consequently, minimizing pressure on biodiversity outside of agricultural fields. Hence, modern agriculture is capable of maximizing the level of biodiversity by reducing the necessary cultivated area, but is minimizing the level of biodiversity in farmers' fields.

Conservation is another interaction and overlapping of biodiversity with agrobiodiversity. In order to conserve wild plant genetic resources, it is essential to conserve their habitat. The in situ conservation method is the most commonly utilized conservation method for wild plant genetic resources. While conserving wild plants in situ (e.g., habitat conservation), there is a spill over effect in terms of wild relatives of crops which may be conserved in these habitats as well; they are, however, neglected by the conservation efforts at present.

Although there are some important similarities, the differences are overwhelming as can be seen in . Plant biodiversity is commonly defined by inter-species diversity as the variation of diversity between species. It is characterized as a large amount of plant species, but a small amount of varieties in each species, whereas agrobiodiversity is determined by a large amount of crop varieties of a given crop, but a small amount of different crop species. Thus, agrobiodiversity is commonly defined as intra-species diversity. Consequently the common method for quantifying biodiversity is the number of species per defined area, whereas the number of varieties per given crop per area defined quantifies the diversity of plants for food and agriculture. Following these common quantification methods, the *centers of high diversity* for biodiversity are tropical forests. Drier ecosystems are, on the whole, far more important for crop resources. The evolutionary process is the dominating force for the development of biodiversity, whereas the evolutionary process is of only minor importance for the development of agrobiodiversity. Breeding activities of farmers over thousands of years, and more recently the scientific breeding, are determining the diversity of agricultural crops. The expansion of domesticated species is associated with human activities migration and trade. Hence, diversity of domesticated species is systematically spread out by humankind. The diversity of wild plants is fixed to their region of origin in general, and is transferred only unsystematically to other sites. If these imported or "exotic" plants are strong enough, they establish themselves at the new place. The extinction of plant genetic resources is commonly characterized by the amount of existing species, whereby the number of varieties per crop species characterizes the extinction rate for PGRFA. Because of the lack of incentives for continuous utilization of traditional varieties, the amount of varieties per crop is decreasing significantly, representing the primary *cause of extinction*. The main cause of extinction of the genetic resources of wild plants is the anthropogen induced changes of habitats (modification, fragmentation, and destruction, see Chap. 2.2).

Because of the high amount of plant genetic resources and the importance of the interspecies linkages, in situ conservation is and will stay the predominant *conservation method* for biodiversity in the future. Consequently, habitat conservation, protected area, and other in situ strategies are aiming at reducing human activities in areas of high diversity. Ex situ conservation is only of importance for working collections. After more than 50 years of intensive collections of PGRFA, ex situ collections are the preferred conservation method for agrobiodiversity, although the importance of in situ conservation is starting to increase, mainly as on-farm management. In contrary to wild plant genetic resources the general in situ conservation strategy to maintain PGRFA is to promote the utilization of traditional varieties by providing some kind of incentives to the farmers. The comparative advantage of in situ conservation can be exploited and the genetic resources maintained in an evolutionary process only by utilizing the old cultivars or landraces.

The externalities may be internalized for biodiversity conservation by making the resource users (or destroyers) account for the environmental damage they cause. In agrobiodiversity there is, however, no legal or moral right to make farmers accountable for the damage they cause by not utilizing farmers' varieties any more, i.e., the *polluter pays principle* is applicable to biodiversity conservation in general, but not to the concepts for agrobiodiversity conservation.

Although the *operational utility value* for genetic resources represents a demand for genetically coded information, at this point of time there are still differences on the operational level: the plant breeding industry requires specific traits, specific functions, e.g., for certain resistance, or certain salt tolerance at a certain development stage of a certain crop.

Table 2.5 Comparison of Biodiversity and Agrobiodiversity

Criteria	Biodiversity	Agrobiodiversity
Definition of diversity:	Species diversity as variation of diversity between species	Genetic diversity as variation of diversity in one species
Common quantification of diversity:	Number of species per defined area	Number of varieties of a given crop per defined area
Centers of diversity:	Majority in tropical humid (forest) areas	Majority in tropical semi and arid areas
Driving force of diversity:	Evolutionary process	Breeding process
Expansion:	Unsystematically by humankind	Systematically by humankind
Extinction characterized by:	Mainly species	Mainly varieties
Main causes of extinction:	Human destruction, fragmentation, modification of habitats	Abandonment of landraces
Prioritized conservation methods:	In situ	Ex situ
General in situ conservation strategy:	Reducing activities of humankind in areas of high diversity (habitat / protected area approach)	Promoting the use of agrobiodiversity by farmers (on-farm management)
Polluter pays principle	Applicable	Non-applicable
Operational utility value:	Pharmaceutically-active compounds (genetically coded information)	Desirable traits (genetically coded function)
Supply for:	All biotechnology industries including plant breeding industries	Conventional and biotechnology plant breeding industries
Involved companies:	Mostly private companies	Public and private companies
Genetic sources for a new product:	Single component	Various components
Research and development system:	Private	Public and private
Transaction chains between collection and ultimate producer:	Short	Long
Revenue-generating potential of products:	High	Low
Redistribution of sales as "benefit sharing" to the providers of genetic resources:	Effective	Ineffective
Genetic resources exchange system:	Bilateral contracts	Multilateral system
Overlap:	*Wild relatives of plant genetic resources for food and agriculture*	

The pharmaceutical industry is also searching for certain functions. But because of advanced and different techniques, they are predominantly interested in the information, which determines the functions. The pharmaceutical and other biotechnology industries mainly synthesize new products, whereas the plant breeders only incorporate genetically coded information into existing organisms. Therefore the utilization of genetically coded functions is still more convenient at present. Hence, biodiversity is the *supply* of genetic resources for all different biotechnology industries, largely involving the private sector and increasingly including the plant breeding industry. Meanwhile, agrobiodiversity solely supplies the conventional and biotechnology plant breeding industry, in which private and public breeding companies are involved. Additionally, most pharmaceutical products derived from plants are based on material from a single *component* (Simpson et al., 1996), whereas in agricultural seed production a great number of genetic inputs are utilized.

In contrast to the situation in the pharmaceutical industry, the users of PGRFA are primarily composed of a huge number of site-specifically operating plant breeding companies and millions of farmers are their clientele. In this system, public agricultural *research and development* plays a decisive role for the improvement of seeds. In spite of the recent trend motivated by biotechnological progress to challenge the status of basic agricultural research on plant productivity as a public concern, public research continues to be of great importance both nationally and internationally. In biotechnology, most of the research is, however, already settled in private companies.

The primary seed products, being the main good of agrobiodiversity, typically pass through several institutions (research institutes and breeding centers) before reaching the ultimate producer, i.e., the seed multiplication companies. These long *transaction chains* differ significantly from the direct utilization in general biodiversity, where the ultimate producer, i.e., the biotechnology industry, is usually already involved in the collection activities. There is a striking difference between the *revenue-generating potential of products* concerning pharmaceutical applications from biodiversity and those involving agricultural products from agrobiodiversity. Because of the low income elasticity of agricultural products, the expenditure share of food has declined over the past 50 years, while expenditures on pharmaceuticals have increased, the rate is even higher than the overall rate of inflation - because of a high willingness to pay for medical advances (Zilberman et al., 1997). Although the potential profits from new drugs can be enormous in the area of pharmaceutics - e.g., 200 million US $ for one drug (Reid, 1995) - this is presently not so much the case for seeds used in agriculture. According to an estimate the total annual global net profit of the world's commercial market for certified seed is less than approximately US $ 2 billion (LEI-DLO, 1994).

Based on the assumption that approximately 4,750 varieties are introduced into the seed market every year on average (UPOV, 1995), the potential profits for one single variety are not comparable with those of pharmaceutics. Based on the amount of annual turnover in the pharmaceutical industry (approximately US $ 235 billion), a *redistribution* of 1% of drug sales derived from plants, for instance, could deliver several million dollars to some source countries under bilateral arrangements (Wright, 1996). Whereas, because of the small annual turnover of the commercial seed industry amounting to approximately US $ 15 billion (LEI-DLO, 1994), a redistribution to the providers of genetic resources would be insignificant. In the pharmaceutical industry competition for access to genetic resources may lead to an *exchange system* of bilateral contracts between large corporations and national governments, whereas the agricultural sector – because of its structure and the high transaction costs entailed in this - appears to be moving towards a further consolidation of the existing multilateral system (Cooper et al., 1994)[13].

As can be seen in Chapter 2.4.2, the wild relatives of PGRFA are midway between the genetic diversity of wild and domesticated plants and may be seen as *overlap*. They are relatives or the original of the domesticated plants, and can be found mainly in the centers of origin of PGRFA. Concerning the conservation possibilities, however, they have to be handled more like genetic resources of wild plants. In situ conservation needs to protect the habitat against human agricultural activities. Consequently, by discussing the conservation of PGRFA, their wild relatives must be treated as a part of the genetic resources of wild plants in general.

Because of the important differences between agrobiodiversity and biodiversity it has been recommended to differentiate between both and to treat agrobiodiversity as defined and biodiversity as a synonym for plant genetic resources excluding PGRFA. Consequently this study will concentrate on the analyses and evaluation of conservation methods of PGRFA, touching only briefly the wild relatives of PGRFA.

2.4 Methods for Conservation of PGRFA

There is, however, a fundamental divergence of views on what objectives should be the focus of the conservation of genetic diversity. Ethical aims oriented at preserving all existing biodiversity stand in opposition to anthropocentric objectives which consider genetic diversity only worth maintaining to the extent that it serves human kind at present or in the future. Conservation policies will be pursued with quite different sets of instruments depending on whether maximal or optimal conservation has been chosen.

The present political discussion, in the "Commission for Genetic Resources for Food and Agriculture" (CGRFA), for instance, is characterized by a mutual agreement on the conservation of the diversity of PGRFA. The question whether to optimize or maximize the conservation is, however, not yet on the agenda. It

[13] The different exchange systems mentioned are discussed in Chap. 4.4.

seems that the representatives of most countries implicitly want to conserve all existing diversity represented by all existing crop varieties still in farmers' fields.

Three fundamental objectives for the conservation of PGRFA can be identified for the time being:

- *Freezing for future utilization* through long-term ex situ conservation conserves PGRFA in their present constellation for future generations to come.
- *Intended adaptation of PGRFA to changing environmental conditions* by long-term in situ conservation exposes genetic resources to ecological pressure enforcing natural changes and permitting continued co-evolutionary development (Vaughan and Chang, 1992).
- *Storage for appropriate and convenient access* supports the supply of genetic resources as raw material. The stored collections need to be easy accessible for on-going breeding programs by scientists, farmers and interactive groups of farmers and scientists.

Especially in botanical gardens, plant collecting and displaying has a long history dating back several hundred years. Without having the explicit objective of conserving plant species, botanical gardens became the first conservation sites for plants and are still an important conservation institution. The terminology *ex situ conservation* is applied to all conservation methods in which the species or varieties are taken out of their traditional ecosystems and are kept in a surrounding managed by humans. Starting with the collecting activities of N.I. Vavilov, until recently, most conservation efforts for agricultural crops have concentrated on ex situ conservation; particularly on seed genebanks. Great emphasis was placed on germplasm collecting during the 1970s and 1980s. As a result, the agricultural plants are presently dominating the ex situ collections by far. Defined as the conservation of plants in their ecosystems *in situ conservation*, has been traditionally used for the conservation of forests and of sites valued for their wildlife or ecosystems (FAO, 1996a). In recent years, however, the need for in situ conservation was emphasized, above all at the United Nations Conference on Environment and Development in 1992 (UNEP, 1994a) and during the preparatory process for the International Technical Conference on PGRFA (FAO, 1996a).

As can be seen, both termini, ex situ as well as in situ conservation, originate from the conservation activities of wild plants and animals. They were adopted for the conservation of PGRFA, but do not fit perfectly. On the one hand, PGRFA can only be maintained by human management, i.e., cultivation in farmers' fields. Consequently, strictly following the terminology, all PGRFA are in a state of ex situ management already. On the other hand, in situ conservation for PGRFA is, sticking to the terminology in the narrow sense, not possible, because domesticated plants do not have per se a natural habitat, and if left alone in the natural habitat of their wild relatives, they will have very little chance of survival. Therefore, the terminology will be applied here in a broader sense, i.e., ex situ conservation is defined as the management of domesticated plants or parts of them, outside of their common surroundings, mainly the farm as the agricultural production unit. Following the broader framework of definition, in situ conservation is defined here as all activities to conserve PGRFA in their common surroundings. Although this is a quite practical definition already, there will still be a scope, where conservation activities cannot be clearly classified. Both conservation methods shall be described and analyzed in the following section.

2.4.1 Ex Situ Conservation Management

Following the three defined objectives for conservation, the ex situ conservation methods must be analyzed; the existing management of conservation must be described in the following section.

The ex situ conservation of plant genetic resources developed largely from earlier efforts by botanists and plant breeders to collect and use crop genetic diversity in their breeding programs. There is a variety of different possibilities to store PGRFA ex situ:

- *seed genebanks*: the conservation of all the major cereals dried to a low moisture content and stored at low temperatures over long periods of time.
- *field genebanks*: for food crops which cannot be conserved as seeds. The plants are grown and managed in fields as living collections of accessions with short up to medium term conservation.
- *in vitro*: for vegetatively-propagated species and species with recalcitrant seeds as sterile plant tissue or plantlets either under slow growth conditions on nutrient gels for short or medium-term conservation or in liquid nitrogen (cryopreservation) for the long-term conservation.

In the literature, in vitro storage is mostly a synonym for slow-growth storage. But cryopreservation is another in vitro conservation method of increasing importance. Cell division and metabolic processes are stopped due to temperatures of minus 196^{0}C, and plant material can be stored without alteration for long periods of time. Cryopreservation is a promising option for the safe, long-term storage of germplasm of those species which can easily be regenerated into whole plants, of vegetatively-propagated species and species with recalcitrant seed (Withers, 1990). Cryopreservation may also be used as a technique for the long-term storage of many orthodox species (Stanwood et al., 1981).

The theoretical differentiation between base collection as the long-term storage, which conserves plant genetic resources for the future and the active collection - as that part of the base collection, which is utilized at present is of significant importance for the analysis of conservation activities. Although this differentiation of different collection types exists in theory, the financial, management and other restrictions do not always enable the genebanks to divide their collections into base and active collections. The majority of genebanks have only one type of collection, which is used for all of the different purposes (WIEWS, 1996).

Fig. 2.6. Interdependency between factors determining ex situ conservation methods

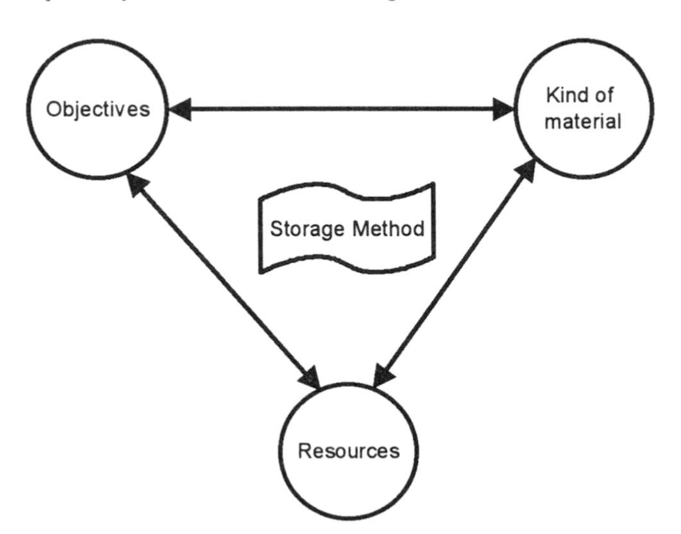

The storage method to choose can be differentiated according to the three determining factors (see Fig. 2.6.):

- The *material* collected determines the potential conservation methods (FAO, 1993) . Depending on the plant species-specific inherent longevity and physiological storage behavior as well as the quality of the collected material, the accessions can be conserved in different ways. In general, the material for conservation can be differentiated into species with orthodox seed[14], vegetatively propagated crops, species with recalcitrant seeds[15], perennial species which produce small amounts of seed, e.g., some forage species, and species with long life cycles (e.g., trees) or intermediate germplasm material.
- The available *resources*, e.g., technologies, financial, institutional, and human resources have a major impact on the decision, which conservation method to use.

---

[14] According to Roberts, orthodox seeds have been defined as: "Seeds for which the viability period increases in a logarithmic manner as one reduces the storage temperature and the moisture content of the same". (Roberts, 1973).

[15] Recalcitrant seeds do not survive under cold storage and/or drying conditions used in conventional ex situ conservation. They can be conserved ex situ for periods of weeks or months only. Important tropical and sub-tropical crop species producing recalcitrant seeds include fruit trees (e.g., mango, jackfruit, citrus, avocado), cash crops (e.g., coffee, tea, cacao, coconut), forest trees (e.g., oaks, maples, *Podocarpus* spp., *Araucaria* spp.), some spices (e.g., cinnamon, nutmeg) (Cromarty et al., 1985).

- Furthermore, the *objectives* and time-frame of conservation determine the conservation method, e.g., base collection, active collections, working collections, community seed collections.

These three factors do not independently influence the choice of conservation method. Rather an interdependency exists between them. Conserving crops with recalcitrant seeds implies either only a very short conservation time of the seed, or utilizing in vitro storage facilities, or conserving the crop in a managed natural environment, e.g., botanical garden. Conservation in botanical gardens is generally more cost-intensive than the seed storage in genebanks (Virchow, 1996b). Therefore, a country might decide not to conserve specific recalcitrant PGRFA, even though they are endangered. On the other hand, there might be a country with the objective of long-term conservation of all their PGRFA. Because of the lack of appropriate technologies this objective is not operational. The lack of human resources may also lead to unsatisfactory storage objectives, because the storage methods are not appropriate for the specific PGRFA to be conserved.

Routine management operations of ex situ conservation include activities as:

- collecting including survey and inventory
- Duplication: for safety reasons the collected accessions should be duplicated twice and stored long-term at other conservation locations in the so-called "black box".
- Regeneration of genebank accessions: regeneration is necessary for accessions of base collections to retain a certain level of seed viability to secure the longevity of the stored seed.
- Multiplication: the need to multiply seed stock is defined by the recommended quantities for base collections and the intensive exchange of specific accessions.
- Documentation management: plant genetic resources can be of any use for plant breeders or other present or future users only with adequate information. For optimal conservation and utilization, the basic data (passport data, characterization data, primary and secondary evaluation data) should be available for each accession.
- Storage: storing the germplasm in adequate storage facilities is the crucial point of the whole conservation management.
- Germplasm health management: the accessions should be screened for pathogens. This is important for the health status of one's own collection as well as for the introduction and exchange of accessions.
- Germplasm movement: it is increasingly important for the genebanks to promote and facilitate the distribution of germplasm mainly to plant breeders, other researchers and to farmers.

Little information is available on the germplasm movement. Only some genebanks and national programs have drawn up some statistics. Over the past three years, for instance, the CGIAR centers have distributed an annual average of over 50,000 accessions to national programs all over the world. (SGRP, 1996). Similarly, between 1992 and 1994, the USA distributed over 100,000 samples each year (FAO, 1996h).

2.4.2 In Situ Conservation

The above-mentioned ex situ conservation methods are increasingly being complemented by efforts to maintain and use the diversity of PGRFA either in its natural habitat (especially wild relatives and forestry species) or in locations where the material has evolved, i.e., on farms or in home gardens. In the latter cases, farmers manage diversity and maintain it through use in their production systems. As defined in the Convention on Biological Diversity, in situ conservation for PGRFA *"... means ... the maintenance and recovery of ... domesticated or cultivated species, in the surroundings where they have developed their distinctive properties.*" (UNEP, 1994a).

Although the ex situ conservation is dominantly utilized for PGRFA, in situ conservation, known as strategy for conservation of ecosystems, habitats and the general inter-species diversity of wildlife therein, has recently entered the stage for conservation of intra-species diversity of PGRFA. This historical development has two implications for PGRFA conservation:

1. The outline for in situ conservation was designed for the conservation of wildlife and later additional for wild flora, without considering the distinctive characteristics of PGRFA conservation, i.e., the conservation of the intra-specific genetic variation of crops and their wild relatives, hence positive conservation impacts for wild or weedy crop relatives occurs as an unplanned result[16](Hoyt, 1988).
2. Because of the historical process, parallel institutional and organizational structures for conservation activities have been developed[17] and links between those concerned are extremely weak in almost all countries. Furthermore, the above-mentioned (Chap. 2.3) competitive relation between conservation of plant genetic resources diversity in general and the diversity of PGRFA conservation dominate the relation. Therefore, PGRFA conservation has not been seen as an integrated part of the general plant genetic resources conservation.

The conceptual framework for in situ conservation for PGRFA is just emerging (see Fig. 2.7.), because only some countries have been involved in in situ conservation programs for PGRFA until now (see Table 2.7). In addition to the conservation of wild relatives of PGRFA, where in situ conservation was incorporated in Biosphere Reserve management objectives (Iltis et al., 1979), or habitat conservation, in situ conservation for PGRFA is conceptualized as on-farm management and improvement for the conservation of landraces and old cultivars. The different in situ conservation methods can be systematized according to the

[16] Hoyt's survey came to the conclusion that even though some protected areas in Europe overlapped with the geographical ranges of the wild crop relatives of apples, plums, cherries, peaches, and almonds, there was very little specific consideration of these genetic resources within the protected areas descriptions or management plans (Hoyt, 1988).

[17] E.g., on the international level the secretariat of CBD is in charge of the conservation of general biodiversity and FAO of the PGRFA conservation.

material conserved (see Fig. 2.7.). On the one hand, there is the conservation of crop wild relatives, mainly a by-product of conservation of general natural areas in kind of - among others - biosphere or natural reserves and habitat conservation. Only seldom and more in recent times, natural reserves are established explicitly for the conservation of crop wild relatives. The other major part of in situ conservation is the on-farm management for the conservation of landraces and old cultivars. Until now programs for on-farm management of PGRFA have been rarely implemented and rarely documented in the scientific literature.

In addition to these designed and implemented programs, on-farm conservation of crop genetic resources is carried out by numerous farmers all over the world. These farmers live in complex, diverse, risk-prone environments, where local livelihoods depend on subsistence farming. They do not have the objective to conserve landraces, but rather to produce food and other agricultural products. The continued use and maintenance of landraces by farmers is carried out for a complex number of anthropological and socio-economic reasons. One major reason is that these farmers are unable to utilize modern varieties for food production due to their socio-economic framework (see Chap. 2.2.2).

The concept of on-farm management and improvement *"provides a mechanism by which the evolutionary systems that are responsible for the generation of variability are conserved*" (Worede, 1992). The level of intra-species diversity in PGRFA is a result of manifold impacts and does not remain static, but rather continues to evolve. Therefore, the concept of on-farm management seeks to enable the processes and surrounding structures, which were responsible for creating the existing variability of landraces, in order to continue to influence the genetic development. The advocates of this concept stress the importance of local systems of knowledge and management, local institutions and social organization as well as several other cultural and socio-economic factors, which determined the diversity development in the past and which will continue to maintain and develop the diversity at present and in the future. Where existing, the in situ activities tend to be more decentralized, more independently organized. Existing on-farm management programs can be categorized into:

- *Targeted approaches*, which prioritize the conservation impact of landraces with significant interest at local, national, regional, or global levels. Furthermore, the increased supply of enhanced seed for breeders and farmers resolving from the activities are also of relevance (Altieri et al., 1987). One of the first "targeted approaches" was initiated in 1988 in Ethiopia. In the drought-prone areas of Welo and Shewa provinces 21 farms were selected for the project, covering sorghum, chickpeas, teff, field peas and maize[18] (Cooper and Cromwell, 1994).

[18] In collaboration with the national genebank, farmers select populations grown in their fields by phenotype. The populations are maintained as distinct from each other, although the system allows for pooling similar landraces and even the introgression of valuable genes from exotic sources (Cooper and Cromwell, 1994).

- *Integrated conservation and development approaches*, which usually link the conservation of landraces and old cultivars with specific values to the cultivation of these varieties directly. These valuable varieties are introduced or reintroduced to a certain agro-ecological region or production systems. Additionally, these valuable varieties are promoted for breeding and adaptation purposes on farmers' sites ("participatory plant breeding" approaches). Breeders have increasingly turned to such sources and turned away from traditional collections, in which variability is stored in a static state (NRC, 1993). Programs with this approach involve often NGOs in "grass roots" PGRFA activities, e.g., the MASIPAG Program (Farmer-Scientist Partnership for the Advancement of Science and Agriculture) (Vicente, 1994).

Fig. 2.7. In situ conservation of PGRFA

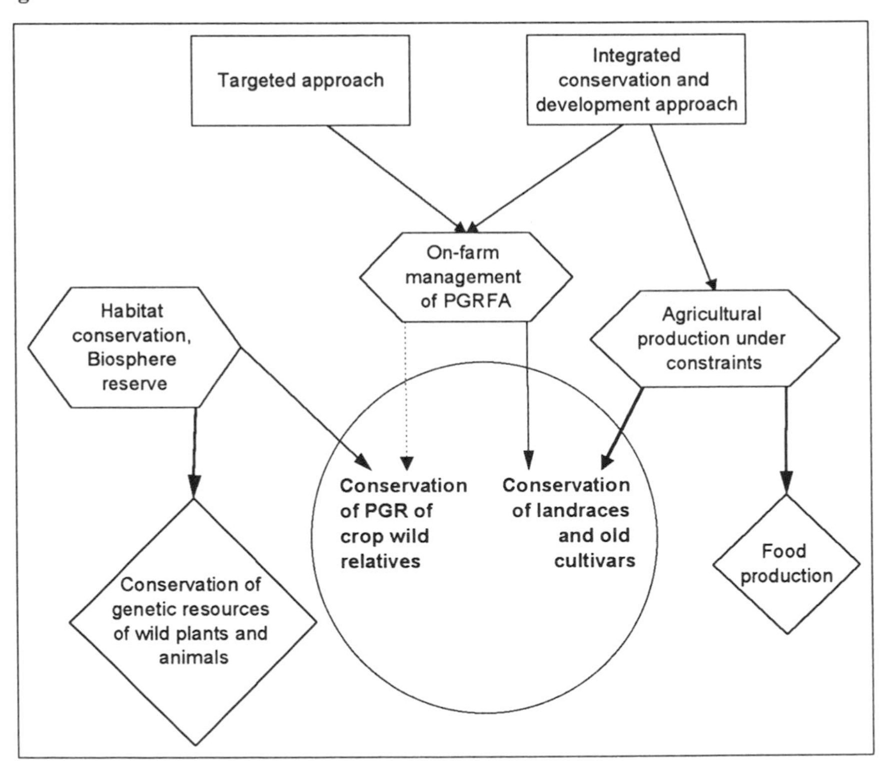

As can be seen from the examples reported (see Fig. 2.7.), in situ conservation for PGRFA is mainly compiled for wild relatives of crops. This indicates that programs set up for the conservation of old cultivars or specific landraces are an exception. Including all "grass-root" activities, promoted by various NGOs, there are more conservation activities relating to old cultivars and landraces than the Country Reports suggest.

Biosphere reserves, characterized as *"... areas of terrestrial and coastal/marine ecosystems, where, through appropriate zoning patterns and management mechanisms, the conservation of ecosystems and their biodiversity is combined with the sustainable use of natural resources for the benefit of local communities, including relevant research, monitoring, education and training activities"* (Robertson, 1992) protected areas, natural reserves are all conservation methods, conceptualized for the in situ conservation of wild plant (and animal) diversity. In some cases, however, a positive impact on the conservation of wild plant genetic resources for local food security is visible, as reported in some Country Reports (see Fig. 2.7.).

Efforts are presently being undertaken to protect areas for the in situ conservation of some wild crop relatives. An important example is the Global Environment Facility (GEF) initiated and funded project to conserve wild crop relatives of cereals, horticultural and ornamental flower crops, medicinal plants and forest trees in Turkey (GEF, 1997). Another important example is the Sierra de Manantlan Biosphere Reserve, which was enlarged by the Mexican government specifically to protect maize and other wild crop relatives. It covers 139,000 hectares, and contains the site where a new species of perennial maize (*Zea diploperennis*) was first reported in 1979.

A majority of countries are presently calling for on-farm conservation activities. They promote these activities, although, there is not very much experience concerning the management and success of this conservation approach as was shown (FAO, 1996a). The majority of the projects cited in the country reports show that of them are integrated conservation and development approaches and are not limited to pure in situ conservation. These programs are usually linked to support for traditional agricultural systems, to crop improvement through participatory approaches to plant breeding, or to community genebanks which are a form of ex situ conservation.

2.4.3 Combination of ex Situ and in Situ Methods

Even though technological development has created various possibilities for ex situ conservation, agrobiodiversity cannot be conserved entirely by laboratory methods (Gupta, 1991). Hence, it seems obvious that none of the methods is able to realize all expected objectives of plant genetic resources conservation; ex situ and in situ conservation activities are not combined very much at present. There is a significant differentiation in the kind of PGRFA stored in ex situ and in situ conservation at present: if technologically possible, the old cultivars and landraces are conserved ex situ, whereas crops without orthodox seed and the wild relatives of crops are mainly conserved in situ.

---

[19] In collaboration with the national genebank, farmers select populations grown in their fields by phenotype. The populations are maintained as distinct from each other, although the

Table 2.6 Countries reporting in situ conservation programs involving PGRFA

	Countries	Dominating in situ conservation programs
Africa	Burkina Faso	Landraces of millet and sorghum
	Ethiopia	Wild relatives of coffee
		Landraces of teff, barley, chick pea, sorghum and faba bean
	Malawi	No detailed information available
Americas	Bolivia	No detailed information available
	Brazil	Wild relatives of cassava, peanut
	Columbia	No detailed information available
	Mexico	Wild relatives of maize
	Peru	No detailed information available
	USA	No detailed information available
	Azerbaijan	Wild fruit trees and shrubs
	Kyrgyzstan	Wild fruit trees and shrubs
Asia	Philippines	No detailed information available
	Sri Lanka	Wild relatives of rice, legumes, spices, wild fruit trees
	Thailand	No detailed information available
	Turkmenistan	Wild fruit trees and shrubs
	Bulgaria	Wild relatives of various crops
Europe	Czech Republic	Wild relatives of various crops
	France	No detailed information available
	Germany	Wild relatives of apples, pears
	Greece	No detailed information available
	Hungary	No detailed information available
	Turkey	Wild relatives of cereals, horticultural and ornamental flower
Near East	Egypt	No detailed information available
	Israel	Wild relatives of wheat

Source: relevant Country Reports

The two different concepts of conservation were differentiated by those who were the major actors in those conservation activities: ex situ conservation - enforced and promoted by governmental and inter-governmental organizations, and the private seed sector – has been managed for the conservation of those crops, which are mainly of interest at global level. In situ conservation was mainly promoted by NGOs, handling regional or local important food crops as well as crops which are not suitable for genebanks.

system allows for pooling similar landraces and even the introgression of valuable genes from exotic sources (Cooper and Cromwell, 1994).

The limitations of ex situ conservation can be summarized as follows:

1. Because of the current state of technology, many important species cannot be stored in seed genebanks, because they do not have orthodox seeds. Consequently, these species are underrepresented in germplasm collections.
2. The ex situ storage methods do not guarantee a long-term conservation without any negative impacts on the diversity of the plant genetic resources. Genetic shift due to insufficient and inappropriate regeneration, storage, health care, and existing capacities, are decreasing the genetic variation, which existed in the original collection sample.
3. Because of the conservation method, ex situ conserved genetic resources are not exposed to natural and artificial pressure, and can therefore not be expected to evolve and adapt to environmental changes.

Although there are not very many studies yet concerning the potential of on-farm management, the limitations known are:

1. If on-farm management is conceptualized by minimizing the area for each variety to be conserved, there is a risk that some allelic diversity will be lost in a limited population.
2. The cultural and socio-economic factors are important for the development of diversity on the local level, but they are related to ecological, social and technological development. Therefore, the future interest of local communities and individuals in conservation activities cannot be taken as granted.
3. The promotion of on-farm management would be pointless without an very long time horizon of at least 50 to 100 years, which increases the risk of every management concept.

In addition to these limitations, the main advantages of on-farm conservation can be described as follows:

1. On-farm management, as a dynamic form of plant genetic resources management, enables the processes of natural and artificial selection to continue. Consequently, despite the loss of some allelic diversity, on-farm management promotes the development of diversity (seen in a perspective of approximately 100 years).
2. It allows the possibility of conserving a large range of potentially interesting alleles.
3. It facilitates research on species in their natural habitats.
4. It assures protection of associated species, which – in spite of no economic value - may contribute to the functioning and long-term productivity of ecosystems.
5. On-farm conservation is especially desirable for crops which do not receive sufficient attention by the formal sector.
6. On-farm management may contribute to an agricultural development while conserving diversity. Linking the management of PGRFA to the improvement of landraces through breeding may be an appropriate strategy for improving farmers' livelihoods in marginal areas as long as there are no economic alternatives.

The links between ex situ conservation and on-farm management has, in the past, been generally limited to the transfer of germplasm samples from the farmers to the ex situ collections. This one-way traffic of information and goods clearly shows the present sub-optimal utilization of the conservation methods. There is far more potential of interaction to the mutual benefit of the whole conservation system, as was pointed out in this chapter. Presently, the importance of an efficient combination of ex situ and in situ conservation is being increasingly recognized (Hawkes, 1991).

The present situation of PGRFA diversity on farmers' fields can be summarized as follows[20]: although PGRFA diversity in situ is decreasing as regards the area cultivated with diverse varieties, the global diversity of PGRFA in general has not yet decreased significantly. There are, however, some regional and crop-specific exceptions, e.g., China's traditional rice varieties have been totally replaced by modern varieties (Kush, 1996).

[20] This conclusion is based on the experience of the preparational process for the International Technical Conference on Plant Genetic Resources and FAO's background report of the state of the world's plant genetic resources for food and agriculture (FAO, 1996a).

3 Economic Framework of Conservation

After giving an overview of the state and development of diversity generally for wild plants and specifically for PGRFA, the conceptual framework has to be outlined. This chapter will concentrate only on economic aspects of PGRFA conservation.

3.1 Conceptual Overview

In Chapter 2 it was argued that the decline in varietal diversity is a result of the economic factors. The individual farmer rarely benefits from the expected value of the production of agrobiodiversity as a public good. Consequently, he or she has little incentive to continue growing the old varieties or landraces, but an incentive to turn to modern varieties when higher income is expected.

Biological scientists argue that almost all genetic resources are potentially valuable and hence should be conserved (e.g., Wilson, 1988). It is assumed that all genetic material has potential value, because the future technologies and environmental conditions are not yet known (McNeely et al., 1990). Consequently, the future value of existing genetic resources cannot be determined at present. This results in the call to conserve a maximum amount of genetic diversity. Additionally, there are arguments defining the value of genetic resources purely from an environmental-ethical point of view (e.g., Shiva, 1991; Busch et al., 1989; Oldfield, 1989).

On the other hand, conservation of genetic resources require financial as well as other resources, which compete with other activities. The competition for resources implies that genetic resources with currently low expected value will not be conserved. Therefore, a ranking for the most important genetic resources is suggested to determine what should be conserved (Brown 1990). Hence, it is necessary to analyze both the costs of diversity conservation as well as the benefits to the extent possible (e.g., Evenson 1993; Wright 1995).

Conventional instruments are of limited use when it comes to assessing environmental goods. The natural resources are essentially different from other economic goods on account of their concern to the public, the irreversibility of extinction, difficulty of substituting them, and their intergenerational existence (Hampicke, 1991). Consequently, the market is only able to capture a fraction of the overall value of these goods.

Besides the attempt to quantify the value of PGRFA, it is important to identify the beneficiaries of PGRFA conservation. Apart from the breeders who are utilizing PGRFA as a substantial input for their breeding activities, farmers are benefiting from the agrobiodiversity through new, improved seed. Furthermore, all consumers are benefiting through stable or decreasing food price development due to lasting yield increases.

So far, often no price is charged for the utilization of PGRFA through breeders. Furthermore, hardly any remuneration is offered, either to farmers who are maintaining PGRFA diversity in their fields, or to the ex situ conservators, who are involved in the first processing of genetically coded information derived from PGRFA. Although PGRFA are a public good, it is assumed that there are three different conservation levels. Following the private benefit expectations, the individual farmer will maintain a specific level of PGRAF diversity in his or her fields, which is different from the optimal PGRFA diversity at a country level. Moreover, there is a different optimal diversity level in a single country than the expected social benefit at an international level.

Because of the different levels of optimal agrobiodiversity to be conserved, there is a need for adequate incentives at the different levels. Specific farmers may have to be provided with incentives by the national government to maintain or even to increase a certain diversity level. In addition to this, countries may have to be stimulated in the international interest to increase the level of conserved agrobiodiversity beyond their optimum[20]. Adequate instruments have to be found for both incentives. Institutional incentives, the internalization of external effects, as well as economic incentives, the compensation for maintaining agrobiodiversity, may be necessary.

3.2 Economic Valuation of Environmental Goods at a National Level

The process of expanding the measure of wealth of a country to include natural and institutional capital as well as human resources besides produced assets is challenging economists as well as being caused by the increased economic interest in natural resources. Especially in developing countries with a high natural capital stock, economic growth in terms of GDP is challenged today by the question whether this growth was achieved by drawing down the stock of national natural resources. To arrive at a nation's total capital stock, the decline of natural resources and the degradation of the environment has to be included, estimating the environmentally adjusted net domestic product (EDP) (Lutz, 1993). This

[20] The indicators for defining a PGRFA-rich ("agrobiodiversity-rich") country are the amount of accessions stored in ex situ collections (more than 10,000 accessions) as well as the fact whether the specific country is part of one of the 8 gene centres. This may be a sufficient classification as long as comprehensive inventories concerning the existence of landraces in situ are still missing.

includes the user costs of exploiting natural resources as well as the social costs of pollution emissions. Besides the natural resources, the national human resources and institutional capital must be taken into account, which will not be discussed here. Since the beginning of the 1990s, national stock accounting has started to include natural capital (e.g., UN, 1993), besides earlier work in the late 1980s (e.g., Repetto et al, 1989). Although the "green national accounts" (Hamilton et al., 1996) are becoming more important in the development policy-making, two major obstacles for achieving valuable results still remain:

- finding the feasible indicators for a country's natural capital and
- finding the corresponding prices or values for each of the indicators.

After giving a short overview of the potential methods of valuation for environmental goods, the framework for economic valuation methods for PGRFA conservation will be discussed more in detail. Besides discussing how to value genetic resources, the question of measuring genetic resources has to be put forward and some solutions are proposed.

3.2.1 General Concept of Assessing Environmental Goods

The natural assets (land, water, genetic resources and air), which contribute to a sustainable development and which must therefore be included in a country's natural stock accounting, can be divided into non-renewable and renewable resources. Renewable resources, which provide a country with an unlimited annuity by sustainable utilization, are agricultural land, forest, fisheries and genetic resources as well as water and air. Non-renewable resources, namely mineral resources and fossil fuels, reduce the overall natural capital stock of a country, when utilized and not reinvested in other national assets (World Bank, 1997).

The concept is straightforward, still the identification of accurate indicators is hampering the easy implementation. The World Bank, for instance, estimates the natural capital at a country level by utilizing only protected areas as an indicator for the national capital of genetic resources (World Bank, 1997). As Thiele (1994) shows for the valuation of the tropical forest, the exploitation of timber is calculated and integrated in a country's natural stock accounting, but seldom the non-timber benefits, like the water storage capacity. Calculating all these other benefits, the national and international interest to conserve the forest would be sufficient for necessary action. Furthermore, there are non-timber forest benefits calculations (see Lampietti et al., 1995), but they do not take air, as a natural asset, into consideration. By calculating the oxygen production of a country's vegetation, mainly forest, the importance of natural assets would, however, significantly increase.

All these examples showthat because of missing or incomplete markets, the value of an environmental good is only partially reflected in its price. Consequently, the total economic value of natural resources has to be defined by breaking down the overall value into its various parts. Depending on the various goods, the components of the total economic value are derived from the use value, which is normally divided into the direct and indirect use value and the option value, as well as from the non-use value, which consists of the bequest value and the existence value[21]. Fig. 3.1 shows the desegregated values of biodiversity as an example for the overall value of natural assets. One immediately obtains an idea of the anticipated benefits derived from the different values and of the decreasing quantifiability and valuability from left to right in the range of value categories depicted in Fig. 3.1. The differentiation of the existing values of biodiversity makes clear why genetic resources conservation should not only be supported because of the use values, but also because of the non-use values (Turner, 1993, Pearce et al., 1990).

Various valuation techniques have to be applied to quantify the different economic values. A number of techniques, which have been developed for assessing the value of public goods in general, have been utilized for the valuation of natural resources as well. Table 3.1 categorizes all relevant valuation techniques according to the type of market they rely on and depending on the kind of behavior of the individuals concerned. Where possible, the asset or parts thereof are valued based on the average return of production as well as on some specific adjustments. These calculations are based on actual market prices and factual or potential behavior of the involved actors. The values of agricultural land, forest and non-renewable resources are estimated on the basis of this technique. In the case of the absence of a conventional market, prices of surrogate markets may be used for valuing the asset. The underlying assumption for the utilization of surrogate markets is the fact that the natural asset is linked to private goods and therefore conclusions from the demand for the private good can be drawn for the natural asset (Endres, 1995). This technique is used quite frequently for valuing the progress of pollution and of biodiversity in general, especially through national parks by the travel cost method. The majority of use values of natural resources are able to be estimated by the techniques based on conventional or surrogate markets whereas the non-use values mainly have to be estimated by utilizing constructed markets. Underlying all the techniques corresponding to surrogate and constructed markets is the willingness of individuals to pay for the environmental good (Braden and Kolstad, 1991).

Missing or incomplete markets exist not only for genetic resources, but also for land and water in several developing countries (Grohs, 1994; Rosegrant et al., 1997). Consequently, these techniques are applied to very different natural resources. There are several approaches for valuing biodiversity. Most methods treat biodiversity as non-marketed goods and services, utilizing either the surrogate or constructed markets, thereby estimating people's willingness to pay. Depending on the different methods used, economists are estimating different values for different parts of biodiversity.

---

[21] See for more detail Chap. 3.2.2

Fig. 3.1. Economic values of biodiversity

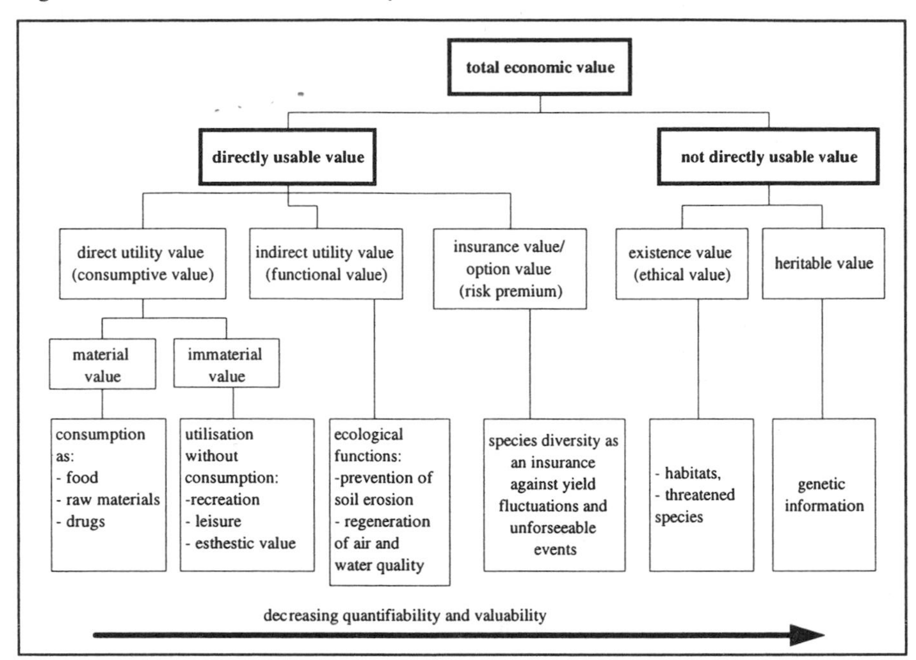

Source: adapted to issues of genetic resources from Munasinghe and Lutz, 1993

Table 3.1. Relevant techniques for the valuation of environmental goods

Kind of behavior:	Conventional market	Surrogate market	Constructed market
Based on actual behavior	Change of productivity loss of earnings defensive expenditure	Travel cost wage differences property values	Artificial market
Based on potential behavior	Replacement cost shadow project (opportunity cost approach)		Contingent valuation

Source: Munasinghe and Lutz, 1993

The concept of *surrogate markets*, which is based on revealed preferences of resource users, is mainly based on the approach used in the travel cost model. The travel cost model is one of the oldest approaches to value natural resources. This method has been widely used for valuing the non-market benefits of outdoor recreation (e.g., Clawson and Knetsch, 1966), especially recreation associated with national parks and public forests (e.g., Bowes and Krutilla, 1989). Other examples for the utilization of the travel cost approach are listed in Appendix 4.

The concept of *constructed markets* aims to get resources users to state their preferences for specific parts of biodiversity. The major approach of this valuation technique is the contingent valuation method. As Carson et al. points out (1995), more than 2,000 studies have been conducted using the contingent valuation method, but comparatively few relate to biodiversity. It has to be pointed out that values estimated from constructed markets are discussed controversially. The main criticism is summarized by Arrow: *"... verbal answers don't hurt the way cash payments do ..."* (Arrow, 1986). Some values estimated using the contingent valuation method are listed in Appendix 4.

Existing market prices based on *conventional markets* can be utilized where an impact on biodiversity effects actual production or productive capability. Some studies, mainly valuing biodiversity by valuing the change in productivity or the loss of earnings by a negative impact on biodiversity, were undertaken and are listed in Appendix 4.

3.2.2 Method for Economic Valuation of PGRFA Conservation

The conservation of PGRFA must be understood as an investment, of which benefits are expected with a long-term delay. Consequently, the appropriate level of investment in conservation activities will be closely determined by the expected benefits and related costs. Economic valuation of PGRFA can contribute to policy and management decisions in guiding allocations of financial resources between PGRFA conservation and alternative endeavors, as well as between various types of PGRFA conservation activities (Artuso, 1994). Additionally, the valuation of PGRFA requires considerations because of the relevance of opportunity cost established by plant genetic resources conservation or the lack thereof. For policy reasons, it is important to differentiate between the private value of plant genetic resources (as the value of the opportunities foregone by the users: its private opportunity cost) and the social opportunity cost (as the value of the opportunities foregone by society as a whole) (von Braun and Virchow, 1997).

As described in general above, the values of natural assets are difficult to estimate. This applies to PGRFA conservation in particular, because the estimation of the non-use values is impractical and the use values are difficult to capture. The intergenerational value of PGRFA and its relevant discount rates are some of the main reasons for the difficulties in estimating the value of genetic resources, besides the uncertainty of the future value of the resources (Pearce et al., 1991). Furthermore, the ecological threshold effects of diversity extinction complicates the valuation (Perrings and Pearce, 1994). Therefore, only a short overview of the valuation methods for PGRFA is given in this chapter.

The total economic value of PGRFA (*TEV*), specified as the overall value of biodiversity, is divided into the direct use value (*DUV*), which is derived from the use of PGRFA in production, consumption and utilization as well as into the non-direct use value (*NDUV*), which arises from the protection of a resource without utilization character:

$$TEV = F(DUV, NDUV) \quad (3.1)$$

The direct use value involves the breeding value (*BV*), which combines the value for breeding, for yield and quality increases, biotic and abiotic stresses, and improvement of ecological functions (i.e., prevention of soil erosion, improving nutrient fixation from alternative sources) as well as the insurance value (*IV*), which reflects the diversity of different varieties of one crop or different crops on farmers' fields as an insurance against yield fluctuations and unforeseeable events[22] .

$$DUV = F(BV, IV) \quad (3.2)$$

The non-direct use value includes the heritable value (*HV*)[23], as the value for the present known and unknown genetically coded information for future utilization as well as the existence value (*EV*), as the value for individuals or groups who do not intend to make use of the resources, but who would feel a "loss" if they were to disappear:

$$NDUV = F(HV, EV) \quad (3.3)$$

Until now, many attempts have been made to estimate the value of various natural asset functions, but not particularly for genetic resources. Conservation of biodiversity is mainly discussed in terms of setting aside state or private land in parks, reserves and forests (Norton-Griffiths and Southey, 1994). Estimates of direct use value of selected wild resources have been made on forest, tourism, animals, and ecosystems in general.

Only few quantitative analyses have been made on parts of the overall economic value of PGRFA conservation. Evenson (1994, 1996b) and Gollin (1996) have contributed to the quantifying of the breeding value by using hedonic trait value estimates.

In general it can be stated that irrigated rice production has increased at 3% per year over the past 25 years. Nearly 60% of that growth is the result of increases in yield from breeding (Hossain, 1994). The situation is similar with wheat and maize: about half of the increase in production has been ascribed to the breeding of new varieties, the remainder is derived from the use of fertilizers, pesticides, fungicides and improved crop management, including irrigation (Riley, 1989).

---

22 The insurance value is often quoted as portfolio value, because the production and yield stability is gained by the maintenance of a wide range, or portfolio, of crops and intra-crop diversity.

23 This value includes the Option Value, as the value of the option to make use of the resource in the future, as well as the Quasi Option Value, as the value of the future information protected by preserving a resource (Arrow et al., 1974; Fisher et al., 1983) and the Bequest Value, as the value of keeping a resource intact for one's heirs (Krutilla, 1967).

Landraces have provided many individual traits which have been introduced into existing improved breeding lines. In this context, Duvick is cited by Salhuana and Smith (1996) that the number of landraces utilized in the five most important maize inbreds are increasing over time: while in 1930, 5 landraces were utilized, it was already 11 in 1960 and went up to 27 in 1990. Evenson (1994), Evenson and Gollin (1996), and Gollin (1996) show with a breeding production function study on rare traits that 10% of the overall gain in rice productivity in South Asia can be attributed to the size and evaluation status of the collections of landraces and related populations used by breeders. This indicates a gain from the material in these collections of about 0.2%, or about US $ 150 - 200 million per year. This value is based on another estimate by Evenson and David (1993). It was estimated that the average value for each of the 1,400 varieties of rice released in India, Pakistan, Bangladesh, Philippines, Thailand, Indonesia, and Brazil in 1990 was US $2.5 million per year. These figures have to be sensitively interpreted, because they represent a gross value before the deduction of costs. Furthermore, as with all other estimates as well, it does not give the isolated value of the genetic material used, but rather the aggregate value of both the genetic resources, as well as the contribution through research inputs as capital, labor and technology involved in plant breeding (NRC, 1993).

Other estimates have been made on the value of CIMMYT-based wheat germplasm to agriculture in OECD countries. These range from $300 million to $11,000 million per year (e.g., Mooney, 1993). The large differences in the estimates reveal the inaccuracy in assessing their value. However, other estimates based on a simple search model show that genetic resources for food and agriculture are not scarce, therefore not of much economic value (Simpson and Sedjo, 1996). These examples show that an economic analysis of the value of PGRFA is still in its infancy. Concentration solely on the consumptive or use value of PGRFA and further research is required to provide better estimates.

Generally speaking, it is assumed that the loss of PGRFA diversity implies the lost chance for the utilization of a specific trait or genetically coded information in crop improvement. The outbreaks of diseases or pests such as the Irish potato famine, which could be given a check by incorporating needed traits into existing varieties serve as justification. But, as Sen has shown (1981), crop failure does not exclusively correspond to famine. It is rather a question of other policy, institutional, and market failures, partly associated with war, violence, or deliberate exploitation. Resistance breeding and utilizing more than only one potato variety were factors rectifying the famine, but not the most important and immediate ones. Taking into account the time lag of 5 or more years between the start of breeding as reaction to a sudden outbreak and the supply of a new variety especially does not verify the assumption that crop improvement through genetically coded information from landraces is the source of insurance against crop losses, as Brown and Goldstein (1984) implied.

Following the estimates of Evenson and Gollin, economic considerations predict an under-investment in the conservation of PGRFA. But the costs of extinction can be overestimated if the opportunities to find substitutes for crop losses are not recognized. On the one hand, crop losses similar to the Irish potato famine will be more likely to occur in developing countries in the future (Smale, 1997), where the immediate food supply through the established food help channels will be faster than breeders' attempt of supplying new varieties. On the other hand, the genetically coded information useful in tackling the problems may be available in other genetic resources or synthetic compounds may be developed (von Braun and Virchow, 1996). There is already some evidence that the elasticity of demand for PGRFA may be much higher than expected: ICRISAT has put its PGRFA collection under the auspices of FAO. Consequently, everyone asking for genetic resources for research and development must confirm that the material received will not be patented or other wise protected and that the benefits resulting from its use will be shared with the country of origin (ICRISAT, n.d.)[24]. Since ICRISAT started sending this standard order form to all requesting germplasm, the overwhelming majority does not send the order form back and seems no longer interested in the germplasm requested (Pundis, 1996). Particularly interesting is that not only private breeders, but also universities and national genebanks are reacting in this way. Because of the uncertainty of the future development in property rights, an agreement like the CGIAR's order form may be more of a disadvantage than charging a specific contribution and being awarded the property rights for the genetically coded information. The willingness to pay for genetically coded information may be higher than the willingness to accept legal uncertainties. This corresponds with statements from the Indian private breeding sector, which is willing to pay for obtaining germplasm from the Indian national genebank (Kush, 1996).

However, as long as PGRFA are existent in the broad diversity, and as long as they are accessible at low costs in ex situ storage facilities, genetically coded information contributes significantly to the breeding value of new varieties. Additionally, conventional market mechanisms are not able to incorporate intergenerational aspects of PGRFA conservation and the irreversibility loss of genetic resources.

At present, the values of the use and non-use aspects of PGRFA are rising simultaneously because of increasing demand on the one hand and decreasing supply on the other. In the future, this value will depend heavily on technical improvements and development, as depicted in and described below.

---

[24] Since 1991, ICRISAT has placed germplasm under the auspices of FAO, based on the Agreement between ICRISAT and FAO dated 26 October 1991. In ICRISAT's Standard Order Form for 'designated germplasm' the person or institution requesting the germplasm has to agree on (ICRISAT, n.d.):
"...not to claim ownership over the material received, nor to seek intellectual property rights over that germplasm or related information, to ensure that any subsequent person or institution to whom they make samples of the germplasm available, is bound by the same provision."

Fig. 3.2. The estimated value of PGRFA over time

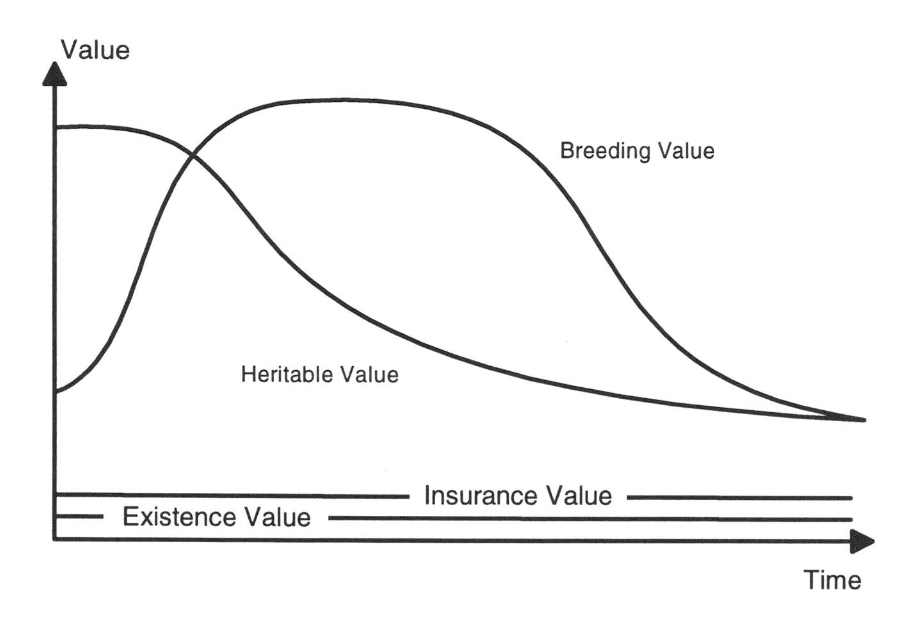

Source: Virchow, 1997a

The breeding value increases in the medium term because of improved technologies and information systems, which simplify their incorporation into existing modern varieties. In the long run, with more improved technologies, the boundaries between species will be overcome and breeders will be able to incorporate genetically coded information from wild plants, micro-organisms and animals into crops, as it is achieved already by introducing insect resistance genes from bacteria into crop plants (Smith and Salhuana, 1996). In addition, as von Braun and Virchow point out (1997), new and improved methods of biotechnology as well as the reproduction of existing and design of new genes will lead to the establishment of a virtual genetic base (e.g., preparation of amino acid sequences) and a combination of genetic information. This point is supported by the prediction of Frankel et al. (1995) that synthetic resistance genes are likely to be developed. In the long run this could curb the demand for genetic resources as a raw material for genetic information, particularly if in vitro reproduction of genetic information becomes cheaper than conservation of genetic resources. This could in turn, lead to the substitution of biotechnological (in vitro) duplicated genes for natural genes and agents. Nevertheless, as long as the in situ and ex situ conservation of genetic resources are still the more cost-effective methods, and assessments of the efficacy of substitutes speculative, the above mentioned potential of substitution will remain confined to a few specific products.

Because of this long-term development the value for plant genetic resources of utilized crops and their wild relatives will decrease significantly. The value will stabilize at a lower level, because local varieties will still be important as a new starting point of breeding. Therefore, the technical development will reduce the intergenerational value of PGRFA. Only the insurance value, defined as actual diversity on farmers' fields, and probably the existence value will remain over time.

Although this development is already predictable today, it is not possible to quantify the time frame for the technical improvement. Hence, the genetically coded information of PGRFA has to be conserved for one or two generations to come.

3.3 Analysis of the Beneficiaries of PGRFA Conservation

The following section is limited to identifying the beneficiaries of PGRFA conservation on the base of the anticipated benefits of PGRFA conservation derived from the different values. For the time being, the best quantifiable value of PGRFA (the breeding value) is passed on indirectly to the beneficiaries of PGRFA, as a positive external effect from mainly marginalized farmers maintaining agrobiodiversity. As will be seen, some of the beneficiaries of PGRFA play a vital role on the demand side, others, however, benefit from PGRFA conservation, but are not directly involved in the demand for genetically coded information.

The *farmers*, resource-poor as well as resource-rich, are cultivating farmers' varieties because of anticipated benefits derived from various values from PGRFA diversity:

- the intra and inter species diversity as an insurance against yield fluctuations and unforeseeable events, balancing considerations of yield maximization and yield stability (insurance value) (Alderman and Paxson, 1992);
- the agronomic adaptation to local conditions and the possibility - and in marginalized areas without access to seed and other input markets the only means - to select seed according to individual and site-specific requirements (breeding value) (Weltzien- Rattunde, 1996);
- the quality, e.g., taste, cooking quality, storability and other qualities, and other material values derived from the direct use of the produced food (consumptive value) (Brush and Meng, 1996); and
- the advantage of farmers' varieties regarding ecological functions, e.g., the prevention of soil erosion (functional value) (Berg et al., 1991).

Reflecting these points, farmers derive private benefits from the different values at the local level. However, the value of the intra and inter species diversity as an insurance usually diminishes with economic and therefore infrastructural and technological development. With the overall development, farmers need to rely less on crop diversity as they are able to offset the effects of crop failure by selling labor (particularly in urban areas) or by drawing upon accumulated assets.

Furthermore, and most important for those with access to markets, farmers may benefit from PGRFA conservation in having the possibility to buy modern varieties' seed, which will enable them to increase or stabilize their income. This seed is partly based on the incorporation of traits or - technically speaking - genetically coded information derived from farmers' varieties. Consequently, the benefit farmers receive for maintaining PGRFA today is the supply of improved and income increasing varieties. Their introduction will, however, result in the loss of farmers' varieties.

The private and public *breeders*, who provide modern varieties, utilize PGRFA as raw material for their breeding activities. Professional breeding as well as farmers' plant breeding are the instruments to convert the present and future needs of the farmers, of the agricultural production processing industry, and of the consumers into new, improved varieties; thereby risking of losing the traditional varieties when farmers replace the traditional varieties with these improved ones. The breeding industry knows about the interaction between genetic resources as raw material for new breeding lines and the displacement of landraces by new varieties, leading to the decline of the amount of genetic resources as raw material in the field for further breeding activities. Therefore it acknowledges the ex situ and in situ conservation as an essential task, particularly to preserve the processes of crop evolution (LeBuanec, 1996). Furthermore, the private sector is interested in the conservation, characterizing, evaluation, and regeneration of PGRFA (Cambolive-Piat, 1996).

According to a survey carried out by WCMC, approximately 7% of the utilized germplasm for breeding is annually incorporated from ex or in situ conserved supply into the existing breeding material (WCMC, 1996). Even though this result seems to overestimate the importance of genetically coded information from landraces, the turnover of landraces in breeding activities is increasing. Smale (1996a) points out that since the late 1970s wheat crosses in a sample of 800 wheats released by breeding programs in developing countries contain an average of one new landrace per year in their pedigrees. Because constant breeding improvements can be made even within a very narrow base (Wych and Rasmusson, 1983), breeders will access other sources of genetic resources only if they are easily accessible and if they are searching for a single specific trait or characteristic. Some studies analyzing the germplasm base of crop species are supporting the idea that breeders have used only a fraction of existing PGRFA while breeding in the past (Fischbeck, 1992; Goodman, 1985). In addition to utilizing the material value for the short- and medium-term perspectives, breeding companies are also interested in the long-term perspective of the heritable value of genetically coded information.

The *biotechnology industry*, as one of the major players on the demand side of biodiversity in general, utilizes genetically coded information originating from PGRFA as raw material for new, genetically engineered plants for agriculture as well as for other products (von Braun and Virchow, 1997). Benefiting by the breeding value, the biotechnology industry is not only improving agronomic interesting traits, but is creating food crop varieties with strong medical impact as well, varieties enriched with micro-nutrients, for example.

Countries involved in the conservation of PGRFA and the international community expect benefits for the national and international breeding activities derived from the raw material as well as from the improvement in the efforts for food security through the insurance value (stabilized or increased food production) and the functional value (reduced degradation of natural resources) as part of the breeding value.

In addition to these beneficiaries, which are benefiting from the conservation of PGRFA directly, the *consumers*, being the main secondary beneficiaries, have to be mentioned as well. Although they do not have a direct role on the demand side, consumers constantly demand more and higher quality food for the same or lower prices. The stability or decrease of consumer prices for agricultural products is due to a complex system of innovations which led to increased agricultural production (Wright, 1996). One of the determining components for this development is the use of PGRFA as raw material for modern varieties.

Summarizing the benefits of the PGRFA diversity and its conservation, the flow of benefits is identified as inter-sectoral (between farmers in marginal areas and high potential areas) and inter-temporal (between farmers presently utilizing modern varieties, which were bred by incorporating genetically coded information from PGRFA diversity, which was maintained in the past). Not only farmers are benefiting from the conservation of PGRFA, but also other actors between the food production and the food consumption. As Wright points out (1996), the social benefits derived from the use of PGRFA, however, are far greater than the profits of the breeding companies. Thus, the major beneficiaries of the use of PGRFA are food consumers in developed and developing countries.

3.4 Incentives for the Conservation of PGRFA

After analyzing the beneficiaries of PGRFA conservation, who represent the demand side of PGRFA, and before identifying the supply side in detail, some conceptional considerations concerning the conservation of PGRFA should be made. After discussing the theory of PGRFA conservation, economic mechanisms of incentives for conservation of PGRFA will be examined.

3.4.1 Theory of the Conservation of PGRFA

As described in Chapter 1, PGRFA, being composed of wild crop relatives and traditional varieties, are supplied in situ by nature and farmers and by local, national and international conservation organization engaged in ex situ conservation as accessions.

Generally, accessions in ex situ storage are freely available to bona fide users upon request; only the transportation costs are charged occasionally (Hammer, 1995). Hence, PGRFA are still handled as free and public good, although - with the adoption of the Convention on Biological Diversity - the free disposal is reduced. Although there is no specific price identified for the supply of PGRFA, there has never been such a huge potential supply of PGRFA than at present. This situation is not the result of increased diversity in farmers' fields, but the improvement of infrastructure has increased the transaction and therefore the availability of PGRFA. On the one hand, the information and transaction infrastructure for the ex situ conserved accessions increases its efficiency, through improved utilization of electronic information systems and technologies for identification of genetic diversity as well as an increased networking of the genebanks. Consequently, the search and exchange costs for bona fide users are presently decreasing. On the other hand, the information of what is still in farmers' fields is increasing in quality and quantity because of on-going inventories and surveys. Therefore, the knowledge of potential PGRFA material is easier to build up and to be widespread.

Even though the current situation gives the impression of a sufficient supply of PGRFA, this does not mean an increase in the renewable part of PGRFA. As has been shown in Chapter 2.2, the total amount of varieties in farmers' fields is decreasing, leading to a reduction in varietal diversity. The loss of traditional varieties is determined, above all, by the production decision at the farm level (see in Chap. 2.2.2). The production decision of a farmer to utilize either the old, traditional varieties or the modern varieties has, after all, an impact on the overall situation of the genetic diversity in situ. This external effect will be thoroughly discussed in detail below.

The farmers' decisions to utilize modern varieties lead to the reduction of land cultivated with specific traditional varieties. If all farmers who cultivated a specific variety replace it with modern varieties, that traditional variety will disappear from farmers' fields. Each farmer will make his or her decision based on their private marginal benefits and marginal costs, which do not reflect the social costs of the variety loss. Depending on the population dynamics of each variety, less than one hectare of cultivated area is, however, sufficient to conserve one variety (Bücken, 1997). Technically speaking, the transformation of land under traditional varieties to modern varieties, characterizing the loss of traditional varieties, is an environmental damage with a very high buffer capacity, determined by the specific ecological threshold effects (Perrings and Pearce, 1994). Consequently, the negative external effect gains importance only with an increasing number of farmers making the same decision. In other words, the environmental damage curve representing the loss of a traditional variety due to land conversion from a traditional variety to modern varieties is increasing by increased land conversion (Swanson et al., 1994).

Farmers are often characterized in the literature as the main load-bearing actors of in situ conservation of PGRFA (e.g., Altieri and Montecinos, 1993). Except for hobby cultivation mainly in industrialized countries, farmers do not, however, maintain PGRFA diversity for its own sake and in accordance with the three objectives of PGRFA conservation. Thus, in situ conservation of PGRFA diversity is a positive externality of the farm activities, based on the farmers' private benefit expectations for other reasons (see Chap. 3.3). Hence, resource-poor as well as resource-rich farmers may produce PGRFA diversity without additional costs.

As long as diversity of PGRFA is not priced[25], and as long as a price for PGRFA is not higher than the agricultural market price, the maintenance of PGRFA diversity in farmers' fields will be negatively correlated to the overall agricultural development in a specific region (as discussed in Chap. 2.2). Therefore, the maintenance of PGRFA diversity in farmers' fields for other reasons than PGRFA conservation may be interpreted as "de facto conservation" of PGRFA diversity (Meng et al., 1997).

Even if farmers are producing PGRFA diversity as a positive external effect to all beneficiaries of PGRFA diversity without costs, the evolution of the marginal costs and benefits of the farmers' production system determine the level of diversity. Farmers conserve PGRFA diversity de facto by utilizing farmers' varieties on their whole farm (mainly resource-poor farmers) or only on parts of it, in spite of access to modern varieties and the agro-ecological possibilities (mainly resource-rich farmers). They do so, because the private anticipated benefit of their production system (including traditional varieties) in terms of insurance, breeding, taste, and so forth, is higher than that of an alternative system (including a higher amount or solely modern varieties), as has been shown in Chapter 3.3 and in different studies, e.g., Brush and Meng (1996). The farmers will go on maintaining their production system and consequently a specific level of PGRFA diversity, as long as their private marginal benefit is higher than their private marginal costs because of the renouncement of higher yields or other benefits determined by the change of the production system. A change of the production system to a "modern" one would inevitably reduce the level of agrobiodiversity. Wherever farmers are able to change their production systems, the costs of maintaining the production system and therefore the de facto conservation of PGRFA diversity have to be reflected as opportunity costs of potential income lost by not planting modern varieties. Being able to change their production system because of market integration, resource-rich farmers reflect their opportunity costs for maintaining their production system and therefore the level of PGRFA diversity (Fig. 3.3.).

As described and seen in Fig. 3.3. , the marginal costs of maintaining PGRFA as traditional varieties are the lowest for the group of economic and ecological marginalized farmers, without any other production possibilities and hence without any opportunity costs of maintaining their production system and therefore the level of PGRFA diversity (represented by the marginal cost curve of the resource-poor farmers MC_{rpF}). The only opportunity costs reflected are either the abandonment of agricultural production due to outmigration from the marginal areas or the change to non-agricultural occupation.

When given realistic alternatives, i.e., by abolishing the marginalized preconditions of production, farmers are not prepared to bear the cost of renouncing short-term profit. Consequently they will change their farming strategies, incorporating modern varieties and dropping farmers' varieties. This means that technical improvement and development at the farm level creates the possibility to choose between traditional varieties and modern varieties and the

[25] With all the institutional implications, e.g., property rights solutions, solved.

introduction of modern varieties may occur. Because of the change of their production system, there will be an inevitable decline in PGRFA diversity in their fields. This decline is characterized by the change of area under traditional varieties from q_{rpF} (resource-poor farmers) to q_{rrF} (resource-rich farmers).

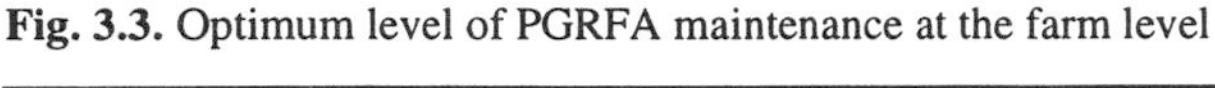

Fig. 3.3. Optimum level of PGRFA maintenance at the farm level

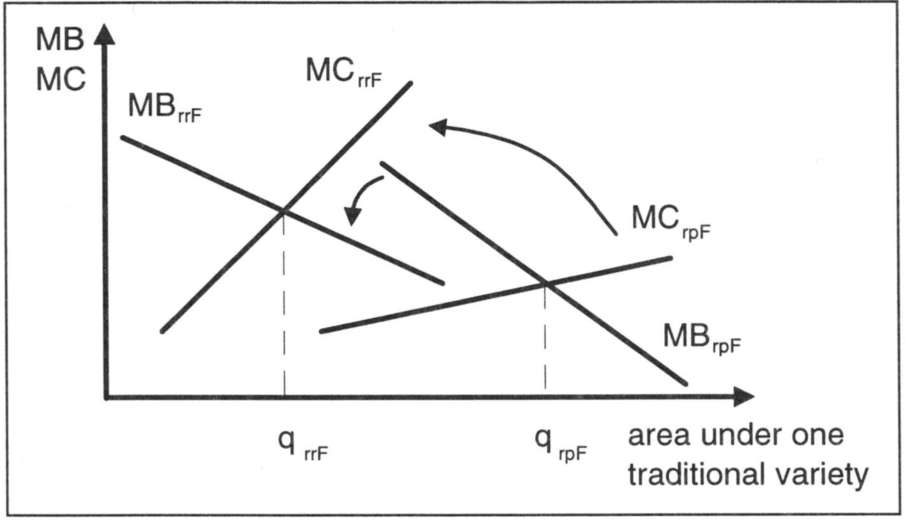

Note: MC: marginal costs; MB: marginal benefit
MCrrF / MBrrF: marginal costs / marginal benefit for resource-rich farmers
MCrpF / MBrpF: marginal costs / marginal benefit for resource-poor farmers

At the farm level, the decision is made whether to cultivate traditional varieties or not, and to what extent. This decision is solely determined by the individual or farm-level benefit-optimizing criteria, which are also influenced by the economic and ecological framework, i.e., the access to resources and technologies. The result being that, with increasing access to technology, the marginal costs of cultivating traditional varieties will increase to the extent to which farmers decide to use modern varieties instead of traditional varieties.

Countries aiming to conserve their PGRFA in situ face one problem by determining their country specific optima of PGRFA diversity level: as discussed above, the farmers provide the country's current PGRFA diversity as free good. Besides the risk that farmers may change their production systems and therefore the level of PGRFA diversity, the country's social marginal costs are mainly derived from the opportunity costs for foregone increased food production through renouncement the utilization of modern varieties. Many countries have to continue and increase the integration of the resource-poor farmers, the "custodian" of PGRFA diversity, into the market to increase national food security (von Braun, 1994b).

As a rule, one traditional variety may be maintained in situ on an area less than $100m^2$ for crops with orthodox seed and less than $250m^2$ for vegetatively propagated crops. For other crop species, e.g., with recalcitrant seeds, perennial species, and species with long life cycles less than ¼ ha are necessary[26] (Bücken, 1997). These requirements are necessary because of the plant population dynamics and in order to keep the genetic variation broad. For safety reasons it may be assumed that each traditional variety may be planted in average on $1{,}000m^2$ and on 10 to 50 farms so that crop failure on one farm does not risk the variety's conservation. Therefore, the total amount of the absolutely vital area necessary to conserve one traditional variety of any crop species would be less than 5 hectares.

Fig. 3.4. PGRFA conservation between private and social benefit

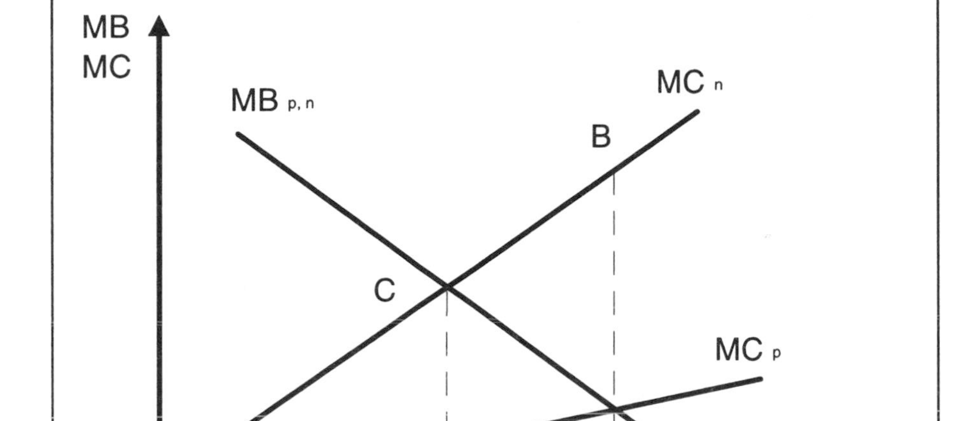

Note: MC: marginal costs; MB: marginal benefit
MC_n / MC_p: marginal costs on national / private level
$MB_{p,n}$: marginal benefit on private and national level

Because of this minimum safety standard (MSS), the country's conservation objective could be fulfilled with the minimum area requirement at q_n in Fig. 3.4. More area under the specific traditional variety will increase the social marginal costs of the in situ conservation in a specific country. In contrast to the country's optimal area need, the farmers - according to their individual optima - utilize far

[26] For instance, the area requirement for maize is $60\ m^2$ and for wheat only $0.75\ m^2$.

more land (characterized as q_p in Fig. 3.4) for the traditional variety than socially adequate reflecting the opportunity costs at national level. Only 74%, 70%, and 57% of the area planted to rice, wheat, and maize in developing countries was planted with modern varieties[27] (Alexandratos, 1995). Pretty (1995) estimates that a total of almost 2 billion people are still not benefiting from modern agriculture. Due to the area still under traditional varieties, where farmers grow the same set or overlapping sets of varieties there is the negative external effect for the society as whole (marked in Fig. 3.4. as area *ABC*). The private decisions may provide optimal levels of on-farm production systems, thereby including a specific level of agrobiodiversity. But it may also provide sub-optimal levels of PGRFA diversity from a social perspective.

Until now, the analysis concentrated mainly on the in situ conservation and the external effects emerging from the farmers' decisions making process. One further important conservation issue must be highlighted below: the positive external effects of a country's ex situ conservation activities at the international level.

Ex situ conservation is done primarily by national and international organizations. The main objective of ex situ conservation is to freeze accessions as carriers of genetically coded information. Taking into consideration that the marginal benefit at the global level (MB_g in Fig. 3.5.) is higher than that at a national level (MB_n) and assuming that the marginal costs for ex situ conservation ($MC_{n,g}$) are equal on both levels, two different optimal conservation levels are evident (see Fig. 3.5.). Individual countries can be expected to implement the necessary measures for securing PGRFA at the respective national equilibrium of q_n. Global initiatives, however, are necessary to ensure the global optimum of agrobiodiversity of q_g in a specific country. This can be done either by compensating countries with the necessary financial resources or by internalizing the benefits. By choosing the compensation mechanism, countries need necessary financial resources at least equivalent to *ABC* in Fig. 3.5. *A* represents a national optimum because beyond this point national economic costs exceed the national economic benefits. If the amount *ABC* is compensated for, then protection of resources at level q_g would also be optimal from the viewpoint of the national economy. In addition to the mechanism of compensation with international transfers (e.g., through international biodiversity funds), internalizing part of the global benefit for national conservators is the other measures discussed for safeguarding the global equilibrium of agrobiodiversity conservation. Hereby, property rights of genetic resources have to be implemented and could serve as a means of market formation. Both mechanisms will be discussed in detail in Chapter 4.

---

[27] As average figures they do not show the differences between countries or regions, so for instance, in sub-Saharan Afica, modern wheat varieties are planted only on 52% of the wheat area and in rice it may be even less.

Fig. 3.5. National and international equilibrium of PGRFA conservation ex situ

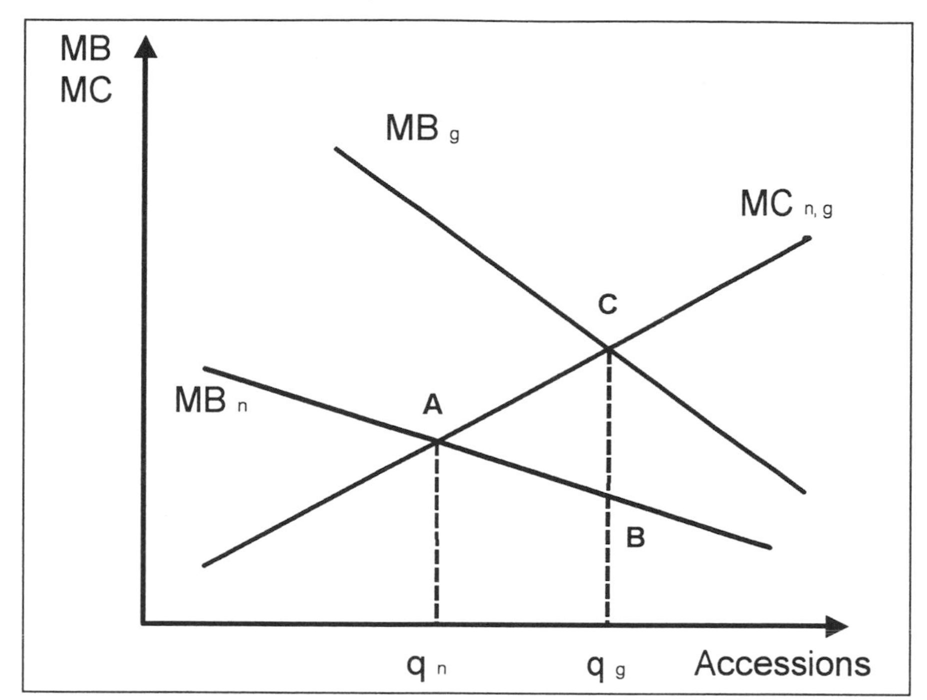

Note: MC: marginal costs; MB: marginal benefit
MBn / MBg: marginal costs on national / global level
MCn, g: marginal benefit on national and global level

The expected and experienced national marginal benefit for the conservation of PGRFA is determined by the state of development in the plant breeding and biotechnology of each country. While countries with an improved capacity in plant breeding and biotechnology potential (e.g., India and Brazil) may capture the social benefit of PGRFA conservation at a national level (q_n), smaller and poorer states - without any or only with basic potential in plant breeding and biotechnology - see few incentives in conserving a high level of diversity (von Braun and Virchow, 1997). Hence, the compensation for these countries must be higher because of lower national optimum. Sometimes countries also are not capable to finance some functioning ex situ conservation facilities to secure PGRFA at their national equilibrium. Furthermore, they are incapable of distributing financial incentives to farmers for the in situ conservation of PGRFA.

Summarizing this section, it can be stressed that agrobiodiversity is mainly produced by farmers as positive external effect without any costs. The farmers are maintaining a level of agrobiodiversity through their farm-specific production system according to the individual optima of the decision-making process at the farm level. From the farm-level perspective, these positive external effects of such activities would need to be identified as clearly as possible so as to permit their efficient internalization, if desired and economically sound. In analyzing the in situ conservation according to the social marginal costs of the country, it has to be realized, however, that the total amount of traditional varieties is produced or maintained by farmers on too much land, which could have the potential of being used for the production of more food by utilizing "modern" production systems, including modern varieties.

Instruments like a tax or coercion will not be effective in reducing the land utilized for less productive production systems, because the farmers in marginal areas have no other choice than to utilize these systems including the traditional varieties. Therefore, the economic and ecological marginalized areas, in which most of agrobiodiversity is produced and conserved de facto until now, needs external investment in infrastructure and technology to reduce the production limitations and to increase national food production. In the long run it should be in the interest of all developing countries, to reduce the economic and ecological marginalisation of areas through investments and at the same time to increase the marginal costs of the farmers utilizing traditional production systems by increasing the opportunity costs of maintaining traditional varieties. Consequently, the amount of area utilized by these systems and therefore the level of agrobiodiversity would decrease.

In the future, assuming an increase in the production in the present marginalized areas, the question will be, with which economic instruments and incentives can agrobiodiversity be kept at the social optimum, securing non-marketable genetic resources? For the time being, the question will, however, remain: how can the marginalized farmer benefit from the produced good "agrobiodiversity" and how can the area utilized for traditional varieties be reduced without losing varieties?

3.4.2 Economic Mechanisms of Incentives for the Conservation of PGRFA

As has been shown, the major cause-complex for the decline of PGRFA in farmers' fields is the missing mechanisms to incorporate the social value of PGRFA conservation into the product of PGRFA. Until now, the price for agrobiodiversity is nothing more than the price for a specific variety on the local food or other agricultural market. This price underrates the value of genetically coded information incorporated in the traditional variety. Because of changes in technology applications as well as economic development, the opportunity costs for the maintenance of a traditional variety will rise, and the incentives to maintain that variety will decrease. Hence, the market forces are presently unable to secure the economic correct balance of conservation of PGRFA and changes in land-use patterns, i.e., the optimal conservation level.

In this section the feasible incentives will be discussed (Chap. 3.4.2.1) and an applicable approach is outlined for a "controlled in situ conservation" (Chap. 3.4.2.2).

3.4.2.1
Feasible Incentives for the in Situ Conservation of PGRFA

As has been discussed, the value of PGRFA cannot be accurately assessed yet (see Chap. 3.2). The main differences in genetic resources to other natural resources are their irreversibility of extinction and the complex feed-back system[28]. In contrast to the induced innovation theory (Hayami and Rutan, 1985; Boserup, 1965) the decline of PGRFA diversity, i.e., the extinction of a specific variety, is not self-correcting. Resource scarcity as well as rising private and social costs, arising from the decline of genetic resources, may induce an increased awareness of the value of genetic diversity. But because of the free-rider situation and the specific feed-back system, it is not self-correcting at the farmer level. Bearing this in mind, the risk of PGRFA decline is the irreversible loss of genetically coded information, if specific varieties were not conserved before their extinction.

When technological and economic changes occur, institutional arrangements have to be implemented to foster PGRFA conservation with increased incentives. Given the appropriate economic incentives, farmers could continue to cultivate traditional varieties and do so for the sake of conservation. Few countries presently provide incentives to farmers to support in situ conservation of their landraces. Proposals for such incentives have, however, been put forward in countries like India, the Philippines, and Tanzania (respective Country Reports).

Incentives attempt to influence the behavior of individuals (farmers) and of whole groups (countries) and may be divided into educational, institutional, and economic incentives. Educational incentives, e.g., awareness promotion, may sensitize farmers to the social importance of the conservation of agrobiodiversity, but as Morris and Heisey (1997) emphasize, profit-motivated farmers will in general not be willing to renounce the additional benefit of a less agrobiodiverse production system to benefit society.

On the one hand institutional incentives are the internalization of the positive external effects by improving the property rights situation for the farmers, which will be discussed in Chapter 4. On the other hand, institutional incentives may force farmers to conserve PGRFA diversity. In addition to being undemocratic, the efficiency of such a method may be questioned, and the enforcement of any sanction is difficult in countries with poor infrastructure.

The problem of enforcing mechanisms internalizing the costs of diversity loss outrules all economic disincentives, e.g., taxes or charges for reducing agrobiodiversity in the field. The specific situation of agrobiodiversity conservation, namely to reduce the area under different traditional varieties to a safety minimum, only permits the utilization of positive incentives. Direct and indirect positive incentives may succeed in using market mechanisms to target individual farmers and to influence their decision making. These incentives should

[28] The breeding value, as main utility value of PGRFA, is utilized mainly by professional breeders. Consequently, the lack of sufficient genetically coded information for breeding will occur first at the breeding level – the farmers, reducing diversity in their fields, will probably receive no signal of value change for a long time, if at all.

be coherent with market mechanisms so as to develop a market for genetically coded information but also to afford protection against the market, i.e., protection against potential external effects on existing, non-use values of agrobiodiversity. The implementation of mechanisms to induce the in situ conservation of PGRFA must be further aimed at the objectives of conservation.

The main direct incentive for selected individual farmers maintaining agrobiodiversity would be a price which compensates the farmer for continuously cultivating a specific variety or maintaining a specific level of agrobiodiversity in his or her fields. Because of the difficulties in measuring the value of farmers' contributions to the conservation of agrobiodiversity, the amount of compensation could be determined by the opportunity costs of foregoing production system conversion to a system with modern varieties. As Gupta (1996) points out, other direct incentives could be awards for diverse production systems or other social based incentives. But the most important non-monetary incentive would be an improved cooperation between farmers and genebanks, especially enabling farmers to receive germplasm from the genebank for further utilization (Gupta, 1996).

3.4.2.2
An Applicable Approach for a "Controlled in Situ Conservation"

The present situation of agrobiodiversity seems to imply that there is no urgent call for action to conserve PGRFA diversity in situ. The agricultural development will, however, continue to take place, therefore the present status quo may change rapidly and unforeseeably in every stage and region. The threat related to such a rapid and uncontrolled change of production systems, due to continuous development and improvement of crop varieties adopted to specific environmental and economic conditions, may be that existing endemic plant genetic resources may be extinguished, before any conservation takes place.

Consequently, a system of a *controlled in situ conservation* should be implemented at the national level[29]. The basic idea is to create an in situ conservation system, which enables the maintenance of an endangered variety on a required areal minimum safety standard (aMSS) and at the same time guarantees a conservation system with the highest possible flexibility based on a self-correcting price as an incentive mechanism. Only if the cultivated area of a specific traditional variety falls under the defined aMSS, i.e., the variety is threatened by extinction, must the instruments of incentives take effect. In addition to the incentive mechanism, the flexibility has to be based on the self-targeting of the farmers cultivating the endangered variety.

The areal minimum safety standard is defined as a minimum of production area needed to prevent any negative effects duo to plant population dynamic issues,

[29] In some literature, in situ conservation is supposed to be reduced to the gene megacenters defined by Zeven and Zhukovsky (1975) (e.g., Hentschel, 1997). Due to the significant differences in landraces and the discussed secondary centers, in situ conservation should be promoted at the country or at least at regional level and not only in the megacenters.

e.g., genetic drift etc. An aMSS can be defined for each crop[30], or on a more general basis depending on other biological criteria (e.g., self- and cross-pollinating plants, generative or vegetative propagated plants). As long as no other measurement for agrobiodiversity is operational, e.g., the promoted "genetically coded information", the in situ conservation of agrobiodiversity must conserve traditional varieties and landraces. By following the rules of the safe minimum standards, it is necessary to conserve all existing varieties in situ [31] (Hohl and Tisdell, 1993; Bishop, 1978; Ciriacy-Wantrup, 1952).

The incentive system has to identify the aMSS for a specific endangered variety, depending on the crop and the varietal specific population dynamics, as well as other risk reduction issues. Knowing the average yield for that variety, the resulting yield quantity can be calculated based on the aMSS. This yield quantity is relevant in fixing the reference price. The approach is discussed in detail in the following section.

The first step is to calculate the conventional gross margin for the expected yield resulting from the aMSS of the endangered variety:

$$Y_{ev} = a_{ev} q_{ev} * P_{ev} \tag{3.4}$$

whereby:

Y_{ev}: *conventional gross margin for the expected yield resulting from the aMSS of the endangered variety;*
a_{ev}: *defined aMSS of the endangered variety in ha;*
q_{ev}: *expected yield in kg per ha of the endangered variety;*
P_{ev}: *market price per kg for the endangered variety.*

The second step is to calculate the conventional gross margin for the average yield of the alternative, improved variety or crop based on the aMSS of the endangered variety:

$$Y_{iv} = m_{ev} * q_{iv} * P_{iv} \tag{3.5}$$

whereby:

Y_{iv}: *conventional gross margin for the expected yield of the improved variety resulting from the aMSS of the endangered variety;*
q_{iv}: *expected yield in kg per ha of the improved variety;*
P_{iv}: *market price per kg for the improved variety.*

Although there may exist non-priced values for the endangered variety, e.g., better processing and storage characteristics, it is necessary to include the opportunity costs into the incentive price. The opportunity costs, due to the foregone gross margin by maintaining a specific level of agrobiodiversity, will be based on the difference between the gross margin of the improved and endangered variety. In this way the foregone benefit for the farmers not cultivating the improved variety can be equalized.

---

[30] For example: the aMSS for one variety of wheat is $0.75m^2$ and for one variety of maize $60m^2$.

[31] The rules of a safe minimum standard can be summarized as follows: unless the benefits associated with the deterioration of an environmental good 'heavily' outweigh the costs of it, no significant deterioration should occur.

The difference between the conventional gross margin of the improved variety and the endangered variety is the opportunity costs (C_{opp}) based on the aMSS for the farmer still cultivating the endangered variety instead of replacing it with the improved variety:

$$C_{opp} = Y_{iv} - Y_{ev} \tag{3.6}$$

The overall incentive for farmers to continue cultivating the endangered variety must be higher than that achieved by adding the opportunity costs to the conventional gross margin for the expected average yield of the endangered variety.

If the opportunity costs are included, the gross margin between the two different production systems are equal; they do not, however, reflect the differences in the production costs. The incentive to cultivate the endangered variety may not be enough, because of the additional risk of flexible prices and consequently flexible benefits for the endangered variety. Consequently, the farmer's calculated gross margin resulting from maintaining the endangered variety will be determined by the farmer's anticipated risk assessment of the price to be expected:

$$P_e = \frac{Y_{ev} + C_{opp}}{a_{ev} * q_{ev}} - R_i \tag{3.7}$$

whereby:

P_e: *an individual farmer's expected gross margin per unit for maintaining the endangered variety;*

R_i: *an individual farmer's anticipated risk-assessment*

Hence, a risk premium must be added as a further incentive. Because of private benefit considerations, farmers will cultivate the endangered variety only if the incentive is equal or higher than the individual farmer's anticipated risk-assessment. The risk premium will be a certain percentage of the conventional gross margin for the average yield of the improved variety or crop based on the aMSS of the endangered variety:

$$R_{ev} = \frac{Y_{ev} + C_{opp}}{100} * r \tag{3.8}$$

whereby:

R_{ev}: *risk premium;*

r: *percentage of the conventional gross margin for the average yield of the improved variety or crop based on the aMSS of the endangered variety.*

Consequently, the gross margin of the demanded quantity of the endangered variety which is offered as incentive (Y_i) for the in situ conservation can be described as follows:

$$Y_i = Y_{ev} + C_{opp} + R_{ev} \tag{3.9}$$

or depicted as in Fig. 3.6.

Fig. 3.6. Calculation elements for the incentives in a "controlled in situ conservation" system

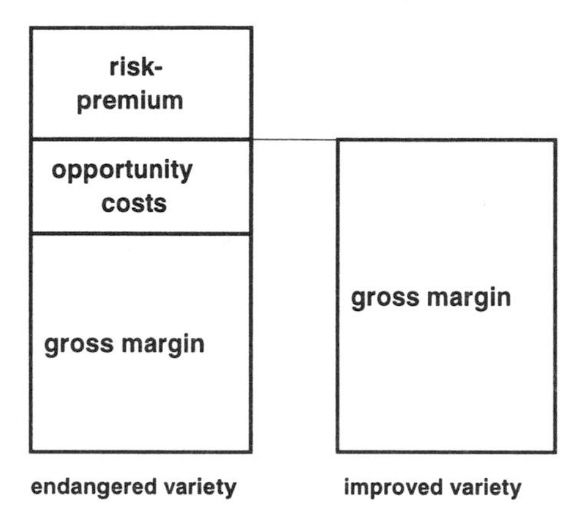

The entire incentive system is based not only on the different prices of the alternative varieties or on the area cultivated but also on the conventional gross margin for the average yield of the endangered and improved variety. This is because the price for a traditional variety may exceed that of a modern variety because of specific advantages or traits, e.g., the Basmati rice with its superior palatability (Alexantratos, 1995). Consequently, the value of a higher yield of a modern variety may be equalized or even surpassed by the higher price for the traditional variety in terms of gross margin. Hence, opportunity costs for maintaining a traditional variety have to be calculated only if the gross margin of the modern variety exceeds that of the traditional variety. Furthermore, it may be assumed that the production costs for a traditional variety may generally be lower than for a modern variety. This will, however, not be integrated into the calculation at this stage. And there still may be some non-market values of the traditional variety, which increases its value for a specific farmer.

The incentive system has to be made operational after fixing the incentive gross margin (Y_i) for the demanded quantity of the endangered variety by the conservator of the demand side. The value, which the demand side is prepared to pay for the whole calculated yield, is defined by the equation described above (3.9). Therefore, the price, being the incentive for one kg yield of the endangered variety, can be deduced as follows:

$$P_i = \frac{Y_{ev} + C_{opp} + R_{ev}}{a_{ev} * q_{ev}} \qquad (3.10)$$

The incentive price P_i will be paid by the demanding conservator for one kg yield of the endangered variety, which is therefore to be in situ conserved. This price P_{i1} will, however, be paid only as long as the supplied quantity equals the expected yield resulting from the aMSS of the endangered variety[32].

The supply will be higher than calculated if more farmers have cultivated the endangered variety, or if the same amount of farmers have cultivated the endangered variety on more land, or the average yield increased. Consequently, the price for one kg yield will be reduced, because the conservator is only interested in his or her fixed reference system.

On the other hand, if the supply is less than expected because of a decrease in cultivated area, the price for each kg yield of the endangered variety will increase. Thereby, the general premises for this self-regulating system will change. Consequently, the basic assumptions for the system must be verified every time and – if necessary – adjusted. The regular time interval for adjusting the system's premises depends on:

- the rate of adjustment of the system's premises;
- the rate of farmers' breeding success for the endangered variety;
- the rate of breeding success for the improved variety;
- the rate of price changes for the endangered variety;
- the rate of price changes for the improved variety;
- the rate of new scientific research results concerning the aMSS for the specific endangered variety.

Because of the self-regulating approach there is one inherent risk in the system: theoretically it may happen that in one year the cultivated area of the endangered variety will fall below the aMSS. This may happen because of an over-supply of the endangered variety in one year, with the consequence of reduced prices for the endangered variety, which will reduce the incentives for farmers to go on cultivating the endangered variety in the next season and the next year. Even assuming the worst scenario that no farmer cultivates the endangered variety for one year or a season and that all seed is therefore lost, this does not mean that the controlled in situ conservation is failing. The conservator demanding the in situ conserved variety will always have seed from the years before to supply the farmers, consequently there is no risk of losing the endangered variety. Additionally, if the regulated price for the endangered variety is not incentive enough for the farmers to continue to cultivate the endangered variety, the demander is always capable of increasing the incentive by increasing the risk premium up to a level resulting in enough cultivated area.

As can be seen, the risk premium is a major element in this self-regulating system. It is an important incentive for a secure in situ conservation of a specific endangered variety in times of a deficiency in cultivated area. The risk premium will encourage enough farmers to cultivate an endangered variety although some alternatives exist. On the other hand, the risk premium serves to encourage enough farmers to supply the yields of an endangered variety to the market site.

The higher the risk premium, the higher will be the number of more farmers averse to risk who will maintain the endangered variety because of private benefit

[32] The supplied quantity is based on the actual area cultivated and the actual average yield.

considerations. Consequently, the area utilized for the cultivation of the endangered variety will increase and will exceed the required aMSS and will therefore increase the social opportunity costs, i.e., the social costs for foregone increased food production by not utilizing the improved variety. The increased social costs will be reflected in lower prices for the endangered variety in the next planting season or year. On the other hand, the lower the risk premium, the lower the number of farmers who will maintain the endangered variety due to private benefit considerations.

In spite of the key position of the incentive price for the endangered variety conserved in situ, the calculations are based on the expected total yield for the defined aMSS. This gives an incentive for farmers to improve their "old" variety, i.e., the endangered variety. Increasing the individual farmer's area productivity, i.e., increasing yield per area, will pay off in terms of higher gross margins for supplying the yield to the demander of genetic resources.

As a side-effect of the conservation of this variety, the improvement of the endangered variety has private benefits for the farmer or group of farmers working with the endangered variety. In addition to receiving higher gross margins by supplying the yield to the demander, the improvement of a variety may have social benefits as well. If the breeding process carried out by farmers results in the change of the genetically coded information of the variety, the farmers will have then produced an additional value, which must be credited as well[33].

The flexibility of the prices, as determined by the annual or seasonal supply of the endangered variety, is necessary for the self-regulating approach of the system. The alternative to a self-regulating system with flexible prices would be either a system handing out incentives to every farmer cultivating the endangered variety by equalizing the differences in benefits of the endangered and improved variety or to construct a fixed system distributing quota to the farmers. The first alternative is a reduction of the incentive for agricultural improvement and development, accepting a level of food production and consequently national food security, which would be lower than the existing potential. The applied opportunity costs of this alternative would be higher than the additional costs arising from the introduction of the risk premium by the self-regulating system.

The second alternative to a self-regulating approach with flexible prices must be analyzed more closely. Assuming that the objective of in situ conservation lies in minimizing the utilized area for endangered variety to the aMSS without risking the extinction of the variety and therewith allowing the highest possible agricultural improvement and development to take place, an administration for distributing and controlling quotas must be constructed or set up. The opportunity costs of this alternative would therefore consist of the administration costs of the system. In addition to these transaction costs, the opportunity costs would be increased by the inflexibility of this system in reacting to changes on the supply side.

---

[33] For more detail about protecting the breeding work of farmers see the discussion on intellectual property rights in Chap. 4.

There are two different ways of targeting the farmers who may benefit from the in situ conservation system:

On the one hand, a certain amount of share certificates may be distributed among the farmers, determining the individual areal minimum or the production level of the endangered variety. This a priori distribution is a more static approach, incapable of reacting flexibly to external conditions or changes and requiring more resources for controlling the system. Additionally, resources to develop decision criteria for the distribution of the share certificates and the implementation thereof are necessary.

Based on a self-regulating price incentive by defining the actual incentive price at the time of supply the a posteriori distribution of benefits for maintaining endangered varieties in situ is more flexible.

Another variation may be to only control the yields of the supplying farmers, paying them a kind of acceptance incentive, based on the opportunity costs plus the risk premium, and leaving the yield to the farmers to use it according to their considerations. This variation would reduce the costs of handling the yield resulting from the aMSS for the demand side. It would, however, increase the costs of the demand side due to increased activities for travelling and controlling.

This approach requires a national centralized system of PGRFA conservation with a strong and central role for the national genebank. It has to be the focal point for all information on ex situ and in situ conservation activities in the country as well as for any collaborations on international level. This information system insures that all relevant information is available even in a country with decentralized structures of conservation and breeding activities. This is one of the major prerequisites for an efficient utilization of PGRFA (see Chap. 5.4). Information concerning the technical development in marginal agrobiodiversity-rich areas must be observed in order to assess the danger of losing distinct varieties[34]. The genebank must be the place for determining what and when controlled in situ conservation activities have to be put in place. The doubling of conservation work can be reduced and the cost effectiveness of PGRFA conservation can be increased only with a centralized information system (see Chap. 5.4).

In this chapter, the conceptual framework and the theoretical issues have been outlined. PGRFA conservation cannot be carried out without reflecting the value of genetic resources. Although an attempt was not made at assessing the value, it is clear that PGRFA have a certain social value. It remains to be seen whether the whole social value can be captured by market prices. A significant breeding value exists as long as the genetically coded information is not substituted by other, synthetically produced information.

Because of this and the other theoretical values discussed, different optima of conservation levels exists. These optima can be reached only through the right incentives, internalization of positive external effects and compensation for conservation efforts. While a very flexible and self-targeting incentive mechanism

---

[34] See Chap. 3.1 for the concept of agrobiodiversity-rich and -poor countries.

is needed for a controlled in situ conservation on national level, a more general approach of compensation may be utilized on a national and an international level.

In addition to crediting the farmers and the countries for their efforts in maintaining agrobiodiversity, institutional development is needed to establish an efficient exchange system for PGRFA. This will be discussed in the following chapter.

4 Institutional Frameworks for the Exchange and Utilization of PGRFA

The major institutional factor determining the threat of PGRFA diversity loss is the lack of or poorly defined property rights for genetic resources. The free good characteristics still existing give rise to the problems associated with the symptom of free rider. Hence, property rights are essential for the conservation of agrobiodiversity: the incentives to maintain or decrease the amount of PGRFA diversity depend on the structure of the various rights of all agents involved. Open access conditions prevail where no property rights exist. Rights must be conferred to avoid the consequences of open access. Reaffirming the sovereignty of nations over their resources, the Convention on Biological Diversity supports governments which modify existing property rights arrangements for genetic resources and build new arrangements.

These institutional changes will affect the economic situation on the informal or formal market for PGRFA exchange. If transaction costs between aggrieved and exploiting parties are prohibitive because of the legal and institutional settings, the exchange system will collapse, reducing the possibility of maintaining agrobiodiversity at a high level.

Systematic survey, collection, and conservation of PGRFA have been under way since the beginning of the century. Today, a complex international and national system for PGRFA conservation is emerging, combining immense energy, human capacity, and financial resources (see Chap. 1.1). In addition to the ex situ conservation activities, PGRFA are still maintained in situ in farmers' fields.

Some major issues are yet to be solved in the ongoing negotiation of the system. Some institutional success has, however, been achieved lately, e.g., the international adoption of the "Global Plan of Action for the Conservation and Sustainable Utilization of Plant Genetic Resources for Food and Agriculture" in Leipzig, June 1996 (FAO, 1996b). The adoption of the global plan marks a highlight in the institutional development of the past 30 years, because it is the first ever global plan formulated and adopted for the conservation of PGRFA.

At the International Technical Conference on Plant Genetic Resources, held in Leipzig, Germany, in 1996, the Global Plan of Action for the Conservation and Sustainable Utilization of Plant Genetic Resources for Food and Agriculture was adopted by 150 countries and 54 organizations. The Global Plan of Action was developed in a country-driven, participatory approach, organized and managed by FAO. The Global Plan of Action was based on the first Report on the State of the World's Plant Genetic Resources for Food and Agriculture. The preparation of this plan and its adoption was also recommended by UNCED in 1992 and the second session of the Conference of the Parties to the Convention on Biological Diversity in 1995. The objectives of the global plan are: (1) to conserve PGRFA as the basis of a worldwide food security; (2) to promote the sustainable utilization of plant genetic resources to reduce hunger and poverty, mainly in developing countries; (3) to promote a "fair and equitable sharing of benefits" resulting from the utilization of plant genetic resources; and (4) to strengthen institutional capacities as well as national, regional, and international programs for conservation and utilization of PGRFA[35].

This chapter will analyze the institutional framework for the exchange of PGRFA, namely the actors in conservation and exchange, the different property systems involved in the process, as well as the different existing and discussed systems of genetic resources exchange. First, the institutional aspects of PGRFA conservation and exchange will be discussed.

4.1 Institutional Aspects of Exchange of PGRFA

The institutional success of adopting a major technical plan of action has not been followed by any breakthrough in the policy debate. The uncontrolled reduction of area under traditional varieties due to technical and economic development, the lack of sufficient resources for incentives at a national level, and the insecurity of additional funds on the international level call for a safe conservation system. The central policy questions concerning PGRFA presently being discussed at the different international fora are:

- the national sovereignty over PGRFA;
- the setup of some kind of property rights associated with genetic resources and its enforcement;
- the arrangement for the access to PGRFA;
- the sharing of benefits of PGRFA between "owners" and users;
- the aspects of financing conservation and the supply of genetic resources, e.g.,in concrete programs.

A conservation and exchange system must solve these problems and integrate all of these issues. Progress has been delayed, because these issues are interrelated, and the main actors in negotiation have contrary objectives. Kloppenburg and Kleinman (1988) identified the declaration of the property rights as the major reason in the deadlock of the discussions ten years ago. The industrialized countries recommend that genetic resources derived from landraces be treated as common heritage but modern varieties have to be protected. The developing countries would also like to declare the newly bred varieties of the seed industry as public good or to protect the landraces instead.

[35] See FAO, 1996b for more detail.

As suppliers of genetically coded information or of the raw material for such information, i.e., germplasm, agrobiodiversity-rich countries are in need for newly developed seed and are therefore interested in its technology[36]. This technology is supplied by countries which are mainly characterized as diversity-poor. Consequently, each side has negotiation resources which are of interest to the other side and the solutions should be feasible (Svarstad, 1994). The negotiation resources are, however, not equally distributed at present, because the industrialized countries (defined as OECD countries for this purpose) have approximately 40% of all conserved accessions at their disposal. Additionally, the accessions of the CGIAR centers are also available for all users. This situation strengthens the negotiation position of the industrialized countries. The exclusion of conserved germplasm collected before the adoption of CBD may serve as an indication for the importance of this fact.

It is, however, not only a dispute between the industrialized countries as agrobiodiversity-poor countries and the developing countries as agrobiodiversity-rich countries, but rather a debate between the suppliers themselves. On the one side the supply of and demand for political lobbying for internalization are determined by existing or expected market power of some countries through property rights solutions, while the prospects of compensation solutions trigger rent-seeking initiatives on the other side (von Braun and Virchow, 1997). Hence, the discussion over the best exchange system is not only driven by the search for the most efficient solution, but also fuelled by country-specific interests.

Furthermore, the system of conservation and exchange of PGRFA which must be created is influenced by other international and national negotiations, treaties, and laws. The network of national and international agreements complicates the negotiations and the potential solutions.

Markets and other exchange mechanism cannot evolve without any international and national framework; or they may evolve only into an informal system without any long-term planning and investment potential. Missing markets and the deficit of other regulation mechanisms are, however, a threat for the conservation and allocation of PGRFA. Hence, markets for the exchange of genetic resources, as part of the internalization mechanism, as well as other regulations to compensate farmers and countries for maintaining a specific level of agrobiodiversity have to be developed.

The present situation is characterized by an imperfect market, causing some specific transaction costs. So far, the whole PGRFA conservation system is based on non-market, i.e., public interactions. The demand side for PGRFA, especially the breeders, the biotechnology industry, and the farmers, has been benefiting from the present system, participating only partially in the costs of conservation and exchange, and above all sharing only a minor part of the existing transaction costs of running the conservation and exchange system. Asymmetric and rudimental information as well as an institutional framework lacking a legal system which defines the individual property rights for PGRFA as well as their

[36] See Chap. 3.1 for the concept of agrobiodiversity-rich and -poor countries.

transfer, are the specific problems hindering the development of a market system with minimal transaction costs for PGRFA conservation and exchange.

With an understanding of why transaction costs play a vital part in the conservation and exchange system of PGRFA, institutional arrangements must be found which present the second-best solution. This solution should meet the Kaldor/Hicks criterion, e.g., the overall PGRFA conservation and exchange system is only then efficient when the additional costs (including the benefit losses) of the establishment and enforcement of the system are less than the additional benefits resulting from the new system.

An efficient system of conservation and exchange of PGRFA has to meet three criteria:

- to provide sufficient incentives at the local, national and international level for the safe conservation of PGRFA,
- to enable a sustainable utilization of PGRFA, mainly in agricultural breeding, and
- to guarantee a "fair and equitable sharing of the benefits" according to CBD and a similar sharing of the burdens of conservation and utilization.

When creating a new conservation and exchange system for PGRFA or improving the existing informal system, one must keep in mind that a system will only be enforced if the resulting benefits exceed the transaction costs of enforcement.

In summary of the discussion concerning the transaction costs for the conservation and exchange of PGRFA, there is an obvious need for an institutional framework with a legal system, guaranteeing property rights and their exchange. Additionally, there is the need for measures to ensure the economic competition as well as the protection of prior social interest on national and global level especially in the form of supportive measures where individual property rights are neither effective nor enforceable.

4.2 Actors in the Conservation and Exchange Activities of Plant Genetic Resources

It is necessary for the institutional framework of an improved or new conservation and exchange system to systematize the main actors in the conservation and utilization process and in a qualitative approach to analyze the anticipated benefits of PGRFA conservation for these actors.

A wide range of different players at a local, national, and international level are maintaining PGRFA, and thereby trying to achieve one or more of the three objectives (freezing for future utilization, intended adaptation, and good accessibility - see Chap. 2.4), in addition to achieving their individual incentives for conservation activities. Rhoades, 1996, groups these players according to their position in regard to access, ownership and compensation (Rhoades, 1996). By grouping the players according to their activities relating to conservation, one can

Fig. 4.1. The pentagon of PGRFA conservators

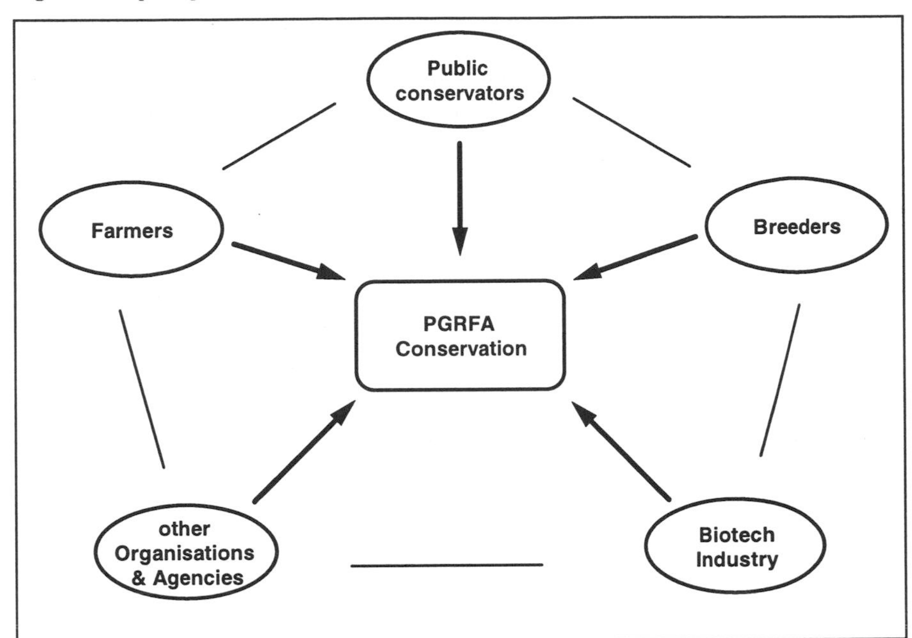

Source: Virchow, 1997b

identify 5 major groups - the pentagon of PGRFA conservators, which is presented below in Fig. 4.1 and discussed in the following section.

As the main players in in situ conservation, farmers play only a small roll in the ex situ conservation, whereas all other actors are mainly involved in the ex situ conservation, with only some isolated activities relating to in situ conservation. The public conservators at the national level dominate the ex situ conservation, storing 83% of all conserved accessions (FAO, 1996a). Hereby, 34% of all accessions are stored in public genebanks of developing countries and 49% in public genebanks of industrialized countries (Iwanga, 1993). According to the FAO survey, 11% of all ex situ conserved accessions are held in regional and international genebanks. The majority of these accessions is stored in the ex situ collections of the Consultative Group on International Agricultural Research (CGIAR). Private breeding companies in industrialized countries store approximately 1% of the accessions and the relevant private companies in developing countries 0.2% (Iwanga, 1993). Finally, it is estimated that less than 0.2% of all ex situ conserved accessions are held by local conservators, i.e., farmers supported by NGOs (FAO, 1996a).

Resource-poor *farmers* mainly utilize non-improved, farmers' varieties. By living in low-potential areas and having less access possibilities to markets or by being confronted with other development restrictions, farmers are being marginalized ecologically and economically and have not benefited as much as others from the diversity of PGRFA through modern varieties (Lipton and

Longhurst, 1989). Resource-poor farmers unable to pay insurance premiums against crop failure, utilize inter- and intra-specific crop diversity as the only option to minimize risk. These farmers constitute over half the world's farmers and produce 15 - 20% of the world's food (Wood and Lenne, 1993). It is estimated that somewhere between 500 and 1,400 million people - approximately somewhere between 270 and 1,000 million in Asia, 160 and 300 million in Africa, and 40 and 100 million in Latin America - are now dependent on resource-poor farming systems in marginal environments (Hazell and Garrett, 1996; Chambers, 1987; Wolf, 1987).

There is still a significant amount of countries in which for one staple crop or the other non-improved varieties constitute at least 75% of the area planted (FAO, 1994). For example, Ethiopia, Eritrea, Burundi and Uganda are still heavily dependent on traditional varieties for their national food production (FAO, 1996a). According to the estimates of Wood and Lenne (1993), 60% of global agriculture depends on the cultivation of traditional varieties. These farmers depend on their own varieties because of the constraints on resource-poor farming systems in marginal environments as well as the difficulties in developing varieties for extreme agro-ecological environments and for existing agricultural practices. Local varieties reveal a significant comparative advantage for these ecological marginalized areas and modern plant breeding has failed to offer more profitable alternatives to farmers' varieties (Ceccarelli, 1993). Regional plant breeding efforts have tended to favor the commercial and semi-commercial sectors, representing the main demand for modern varieties, thereby neglecting the subsistence farmers (FAO, 1996a). Seeds saved by farmers do not necessarily imply that they are only traditional, farmers' varieties. They may also be derived from modern varieties, however are mixed up with farmers' varieties and modified by the farmers over time (Cromwell et al., 1992).

On the other hand, there are the resource-rich farmers who have access to markets, e.g., input, product, capital markets. The majority of these farmers are living in and making use of high potential areas. These farmers significantly utilize to a part modern varieties and are able to choose between farmers' and modern varieties, thereby utilizing the most beneficial varieties for their individual and site-specific objectives (see Chap. 3.3).

In recent years, the production costs for modern varieties may have increased in some regions or countries because of changes in agricultural policies. By reducing or withdrawing subsidies for fertilizers or seed of modern varieties, national policies lead to an increase in the cultivation of traditional crops and varieties. In Zambia, for instance, the area planted with modern varieties of maize, wheat and cotton has decreased, whereby the cultivation of traditional crops has increased, because of the liberalization of the economy since 1991 (FAO, 1996a).

Farmers conserve de facto agrobiodiversity in situ and also contribute to ex situ conservation by supplying the genetic resources to collectors for genebank conservation. *Public conservators* can be identified at a national as well as at an international level.

What was once a scientific response to the danger of losing plant genetic resources has meanwhile emerged into an institutional framework, creating national programs for the conservation of PGRFA in many countries. According to information from FAO, 1996b and FAO's World Information and Early Warning System, 88 countries are in the process of establishing or maintaining such a national program. The nucleus of 95% of all programs is an ex situ collection of national PGRFA (FAO, 1996a). The ex situ activities of 64 other countries without a national program are organized by some research institute, e.g., Russia (FAO, 1996h), or are incorporated into some regional network, e.g., Bangladesh and Norway (FAO, 1996h).

Since the early 1970s, regional genebanks were established in various regions. Some of them, like the Nordic Gene Bank (NGB) with its headquarters in Sweden, hold the base collections for their member countries. In the case of NGB, the member countries do not hold individual national collections, with the exception of Sweden. The Southern African Development Community-Plant Genetic Resources Center (SPGRC) in Zambia stores the duplicates of national collections for Southern Africa while the member countries hold active collections. The regional genebank for Central America and Mexico is based on an existing ex situ plant genetic resources collection established in 1942 (FAO, 1996a).

The primary objective of these regional genebanks is to conserve indigenous plant genetic resources within the region, and thereby to provide training and promote germplasm collection, characterization, documentation and utilization. In order to achieve the objective, a regional plant genetic resources center must be established and a network with plant genetic resources centers in each member state must maintain the cooperation. The funding is provided partially by the member states and partially through financial and technical assistance from other countries or international organizations. The regional genebank for Central America and Mexico, for instance, is financially supported by the German Ministry for Economic Cooperation and Development (BMZ) and supported by GTZ (German agency for technical cooperation) (FAO, 1996h).

Other forms of regional cooperation in conservation activities are the regional PGRFA networks and associated crop networks. These networks, of which most of the regions have one or more, work with collaborative programs aiming to ensure the long-term conservation and increased use of plant genetic resources. They operate through specific working groups in which, curators and breeders from different countries, work together to analyze needs and set priorities for the crop concerned (FAO, 1996a). These networks facilitate the dissemination of information and various collaborative initiatives and they contribute to the development of overall conservation strategies for the ecosystems to which the represented species belong. Furthermore, networks aim to develop locally adapted, improved crop varieties, in close cooperation with the respective international agricultural research institutions.

In addition to the regional genebanks and networks, there are some more important actors on regional and international level. Inter-governmental organizations, such as FAO, UNDP and UNEP allocate financial and technical resources to support conservation and utilization activities at all different levels. FAO, for instance, has been supporting conservation efforts since the late 1950s. In 1957 FAO started a newsletter on plant genetic resources, and since 1961, has convened a series of technical meetings, conferences as well as the regular and extraordinary sessions of the Commission on Genetic Resources for Food and Agriculture. In addition to the technical support, specialized UN trust funds such as the International Fund for Agricultural Development (IFAD) and the Global Environment Facility, international and regional development banks such as the World Bank promote the work of the different PGRFA conservators. Additionally, there are networks which try to co-operate the work of national conservators and to share the results.

Finally, the Consultative Group on International Agricultural Research (CGIAR) represents the main international conservator as regards stored accessions. Genebanks in 12 of their centers conserve 593,367 accessions mainly of their mandate crops (SGRP, 1996). These collections, now being partially placed under the auspices of FAO, represent major global collections for the respective crops.

While farmers have been breeding and are continuing to breed by mass selection from their local varieties to improve their crops, professional plant *breeders*, working in private companies or public-funded research institutes, utilize more complex and expensive techniques to incorporate favorable traits of varieties from different locations into modern varieties. Above all breeders use their own breeding lines, developed out of their working collections of plant genetic resources over ten to fifty years of research (Kush, 1996). Even though constant improvements can be made with the breeders' 'elite' germplasm (Wych et al., 1983), plant breeders are - beside other genebanks - the most frequent users of PGRFA stored in public genebanks (Hammer, 1995).

According to a recent survey of the International Association of Plant Breeders (ASSINSEL), which represents more than 1,000 private breeding companies from 26 countries, 88% of their members are maintaining genetic resources in their own genebanks (ASSINSEL, 1997).

In addition to the national public and private breeding involvement and the multi-national breeding companies, there is an important breeding engagement by the international community through the international agricultural research centers (IARCs). While being the spearhead in the breeding success of the green revolution in the 1960s and 1970s, the IARCs may have to play a crucial role in the next step forward in breeding for sub-tropical and tropical agro-ecological zones. Their role as conservators has been described with the public conservators.

Because of the technological improvement and therewith expanded applications of biotechnology in agriculture, the demand for genetic resources has increased significantly (von Braun and Virchow, 1996). The growing demand for PGRFA is represented in the increasing number of new releases of transgenic plants (de Kathen, 1996). The urgent need to improve varieties by incorporating more genetic diversity into the varieties has been frequently stated (Gustafson, 1997). Beside the many ongoing conservation activities, the *biotechnology industry* is one of the most important actors on the demand side, transforming genetic resources into high value products and determining the increasing importance of genetic resources as an agricultural production factor and therewith its conservation. Furthermore, the biotechnology industry contributes to an increase in the intra-species; in the future it will probably contribute to an increase in the inter-species diversity - as all breeders do by releasing new varieties and new crops.

The private seed sector, conventional and biotechnology industry, has realized its reactive mode in the negotiations concerning the genetic resources conservation in the past and changed to a more pro-active role to represent their interest on international fora by increasing their presence at and direct input to UN meetings (Dickson, 1996).

Other organizations and agencies are also involved in the conservation of PGRFA. In many developing as well as industrialized countries, non-governmental organizations (NGOs), jointly with local people, are actively conserving local plant genetic resources. For example, NGOs promote ex situ conservation and utilization activities on the local or community level. Communities and groups of farmers receive support in collecting their own varieties and storing them in community-based genebanks, which are accessible at any time. The MS Swaminathan Foundation in Madras/India, for instance, manages a community based genebank for farmers in Tamil Nadu and Karnataka, India. Local farmers in Zimbabwe are encouraged to conserve the farmers' sorghum varieties (FAO, 1996h). Furthermore, NGOs have stimulated a conservation competition between farmers in the genetic resources rich Andean Regions of South America by utilizing traditional rural fairs at harvest times (Tapia et al., 1993). Or farmers are reintroducing traditional varieties or crops, because of changes in the economic framework, e.g., the reintroduction of crotalaria as green manure in East Zaire because of the lacking access to the fertilizer market. But also in industrialized countries groups are engaged in the conservation of the existing agrobiodiversity (Vellvé, 1992, Dahl et al., 1992). Besides these grass root organizations, NGOs on the national and international level draw attention to conservation problems, taking part in the international political discussion and negotiation process, e.g., the Canadian Rural Advancement Foundation International (RAFI) or the Spanish Genetic Resources Action International (GRAIN).

NGOs play an increased important role in the negotiation process. Even though their points of view over the public needs diverge from those of representatives from the individual countries, their efforts have pushed the institutional process in the past.

In addition to the above mentioned groups, including the backstopping organizations and institutions behind them, e.g., politicians, genebank curators, which contribute to the conservation and utilization in an active way by maintaining PGRFA in situ or ex situ, some important international bodies exist: the Convention on Biological Diversity (CBD), the Commission on Genetic Resources for Food and Agriculture (CGRFA), hosted by FAO, the Trade-Related Aspects of Intellectual Property Rights (TRIPS) under the World Trade Organization (WTO), and the International Union for the Protection of New Varieties of Plants (UPOV).

Directly or indirectly, these legal bodies influence policy decisions concerning the conservation and utilization of PGRFA for all member countries through international binding and non-binding agreements. This is to be discussed in the next section.

4.3 Protection of Intellectual Property Rights as Enforcement of the Exchange of PGRFA

The specific situation of PGRFA makes it necessary to discuss the exchange system of PGRFA on the one side and the issue of property rights of PGRFA on the other side. Genetic resources have, until recently, been regarded as the *"common heritage of mankind"* (UNEP, 1994a), which was specified in the CBD, according to which the contracting parties reaffirm that the *"... states have sovereign rights over their own biological resources ..."* (UNEP, 1994a, Preamble). This reaffirmed national sovereignty does not, however, solve the problem or propose any solution for the question of ownership of genetic resources.

As discussed in Chapter 4.2, legal bodies influence policy decisions concerning the conservation and utilization of PGRFA for all member countries through adopted conventions and laws. The agreements have an influence on a property rights solution for in situ and ex situ conserved genetic resources as well as the exchange of genetic resources and products developed from their utilization.

4.3.1 International Agreements and the Ownership of Genetic Resources

Meanwhile, so-called unimproved genetic material (wild species and traditional varieties cultivated by farmers) are treated as a freely available resource in one of two systems, while the other is governed by intellectual property rights, i.e., patents and plant breeders titles to newly bred plant varieties produced by conventional breeders and biotechnology companies. Because of the increasing trend towards the privatization and commercialization of research into genetic resources, mainly in the biotechnology industry, there is a high demand for efficient protection systems over the inventions (ODI, 1993).

On the one hand there has been the development of national and international trade and commercial arrangements, determining property rights to genetically and biochemically improved resources. The enactment of the US American "Plant Patent Act" in 1930 was a starting signal, which permitted patent coverage for asexually reproduced plants. Some European countries followed by establishing some variety protection of titles to sexually reproduced plants in the 1940s. In 1970, the USA established the "Plant Variety Protection Act". (Reid et al., 1993).

On an international level, the "International Union for the Protection of New Varieties of Plants" (UPOV[37]) was established in 1961 with the purpose of regulating international trade of protected varieties and ensuring that the member states acknowledge the achievements of breeders of new plant varieties by granting an exclusive property right (Hardon et al., 1994). Regulations under patent law have successively been extended to living organism and consequently entering areas of agricultural interest for approximately 20 years. In the beginning, patents for processing were granted, but in 1985, a genetically modified maize plant was patented (ODI, 1993). The Trade-Related Aspects of Intellectual Property Rights (TRIPS) was enforced at the international level to harmonize protection laws for all technological inventions, including the patenting of plants. Hence, two systems protecting new knowledge presently exist, which advocate the demand side of genetic resources.

As an intergovernmental organization, UPOV was established by the International Convention for the Protection of New Varieties of Plants in 1961. The UPOV Convention was signed in 1961, and revised in 1972, 1978, and 1991. In addition to the positive incentives in the form of exclusive rights for breeders, a further cornerstone of the UPOV's system was the "Breeder's Exemption" - the free availability of protected varieties for the research and development of new varieties. In addition to the exemption for breeders, UPOV convention enables countries to make a further exception from the plant protection rights as regards the seed production and its use on the same farm (farmer's privilege). This privilege was implicit under UPOV '78, it must be specifically defined in national legislation under UPOV '91. The plant protection right does not, however, apply to subsistence farmers or amateur gardeners in UPOV '91. (UPOV, 1995; UPOV, 1992); 26 of the current 32 member states countries are in the process of implementing UPOV '91 at national level[38] (FAO, 1997a).

Beside UPOV being a plant variety protection system, which is legally binding for its mainly industrialized member states, the TRIPS Agreement[39] is based on the principle of providing patents for product as well as process inventions in all fields (OECD, 1996a). Under Section 5, Article 27.3b of the TRIPS Agreement, however, states are allowed to make exemptions from the patenting of plants, if an effective sui-generis protection system for plant varieties also exists. The sui-generis protection system has to be a legally enforceable right either to exclude others or to obtain a compensation for the utilization, i.e., it must be an Intellectual Property Right (Leskien and Flitner, 1997). The TRIPS Agreement, paving the way for the World Trade Organization (WTO), came into effect in 1995. The date

[37] Union Internationale Pour La Protection Des Obtentions Vegetales.

[38] Members of UPOV are: Argentina, Australia, Austria, Belgium, Canada, Chile, Czech Republic, Colombia, Denmark, Finland, France, Germany, Hungary, Ireland, Israel, Italy, Japan, Netherlands, New Zealand, Norway, Paraguay, Poland, Portugal, Slovakia, South Africa, Spain, Sweden, Switzerland, Ukraine, United Kingdom, USA, Uruguay. Countries like: Belarus, Bolivia, Brazil, Bulgaria, China, Ecuador, Kenya, Morocco, Panama, the Republic of Moldavia, the Russian Federation, Trinidad, and Tobago have initiated a procedure for accessions to UPOV, i.e., submitted their laws to the Council. (FAO, 1997a).

[39] Agreement on Trade-Related Aspects of Intellectual Property Rights.

of application, however, depends on the state of the country. Between 1996 and at latest in 2005 all WTO member countries must apply the provisions of the TRIPS Agreement (Art. 65 and 66 TRIPS).

Although approximately 70 developing countries have signed the TRIPS agreement, many of them do not have the appropriate legislation and institutional capacity to regulate the protection of varieties, the import, handling, use, and sell (GCA, 1996, p.12). The protection of intellectual property rights for agricultural innovation is very diverse but generally poorly established in developing countries.

For instance, Brazil excludes from patent law plant and animal species and the processes used for their production as well as micro-organisms and microbial processes. Biotechnology patents are confined to special applications of micro-organisms. Plants and animals are not patentable. The Indian patent law of 1970 excludes agricultural products, as does the Chinese patent law of 1993 (van Wijk, Cohen, Komen, 1993)

The *International Undertaking on Plant Genetic Resources* (IUPGR) was adopted in 1983 strengthening the rights of the suppliers of genetic resources and acting as an counterbalance to the increasing protection of technology resolving out of genetic resources. This undertaking ruled that all genetic resources, including the breeding results of the private plant breeding industry, would be regarded as freely accessible, as stated in Article 1 of the International Undertaking: *"This Undertaking is based on the universally accepted principle that plant genetic resources are a heritage of mankind and consequently should be available without restriction."* (FAO, 1993b, Article 1). Although it is a non-binding agreement, the International Undertaking was not adopted by consensus because some of the industrialized countries with a well-developed seed industry formally had some reservations. In 1987, some annexes including the Farmers' Rights were agreed upon (FAO, 1989). The undertaking accepted the legitimacy of protecting varieties by plant breeders in return for the acknowledgement of the concept of Farmers' Rights, which should be implemented through an international fund. As of May 1997, 111 countries had adhered to the International Undertaking with some exceptions - Brazil, Canada, China, Japan, Malaysia and the USA. While industrialized countries fear the protection of PGRFA in some kind or the other (main point of criticism is the concept of Farmers' Rights for countries like the USA and Canada), some countries like Brazil and Malaysia would like to reduce the scope of the International Undertaking to a few crop varieties in order to keep as many plant species as possible free for other exchange agreements, speculating on higher national benefits. These countries do have a high level of general biodiversity, but a lower level of agrobiodiversity.

The IUPGR is a non-binding agreement assuring the conservation, use and availability of PGRFA by providing a framework recognizing the past, present and future contributions of farmers to the maintenance, improvement, and provision of PGRFA (Farmers' Rights) (FAO, 1993). Sharing the benefits of PGRFA is seen as a crucial instrument for securing global PGRFA. General compensation by the users of PGRFA, breeders and industrialized countries, as well as the internalization of benefits through some kind of joint property rights are the pillar of the concept of Farmers' Rights (Esquinas-Alcázar, 1996). The concept of Farmers' Rights must be understood more as an expression of the principle of global responsibility to conserve landraces and in recognition of the farmers' contribution over time than a legally binding document (Hardon et al., 1994). The implementation of Farmers' Rights, however, is hampered by the lack of international funds for a general compensation approach as well as by the delay in the negotiations of joint property rights. IUPGR was negotiated and agreed upon by the former *Commission on Plant Genetic Resources for Food and Agriculture* (CPGRFA), as an intergovernmental global forum hosted by FAO. CPGRFA was established in 1983 and broadened its scope in 1995 to include other sectors of genetic resources for food and agriculture, beginning with livestock. As part of FAO's *Global System for the Conservation and Utilization of PGRFA* (see Chap. 4.4), CPGRFA consisted of 138 members in August 1995.

At the legal binding Convention on Biological Diversity (CBD) became effective in December 1993, recognizing the considerations of equity and shared responsibility for genetic resources conservation, a number of challenging implications on the *International Undertaking* emerged. These implications relate to the consideration of ownership and accessibility of PGRFA in existing ex situ collections, the realization and implementation of Farmers' Rights, and the establishment of a global ex situ collection network (FAO, 1997b). Consequently, IUPGR must be renegotiated by CGRFA, with the objective to harmonize it with CBD's standard.

4.3.2
International Agreements and Their Coherence

Although CBD and IUPGR advocate the suppliers of genetic resources in their objective of benefit sharing, TRIPS and UPOV generally support the demand side - the breeders and biotechnology industry. Because of differences in the alignment, the lack in coherence between the different agreements has emerged, complicating the development and adoption of an exchange system for PGRFA (see Fig. 4.2)

UPOV is taking a strong position in the breeder's exemption, while weakening the farmer's privilege. A prerequisite for future successful breeding is the free availability of plant germplasm for further research and development. This existing UPOV principle is doubly contradictory: it demands free access to PGRFA in situ and ex situ as well as the free and unrestricted availability of protected varieties for further research and development. Consequently, UPOV's principle offends the Farmers' Rights of the International Undertaking as well as CBD's country's sovereignty over genetic resources in general. On the other hand, UPOV's objective of free access of protected varieties for research and development conflicts with TRIPS' patent system.

Fig. 4.2. Emerging conflicts between international agreements

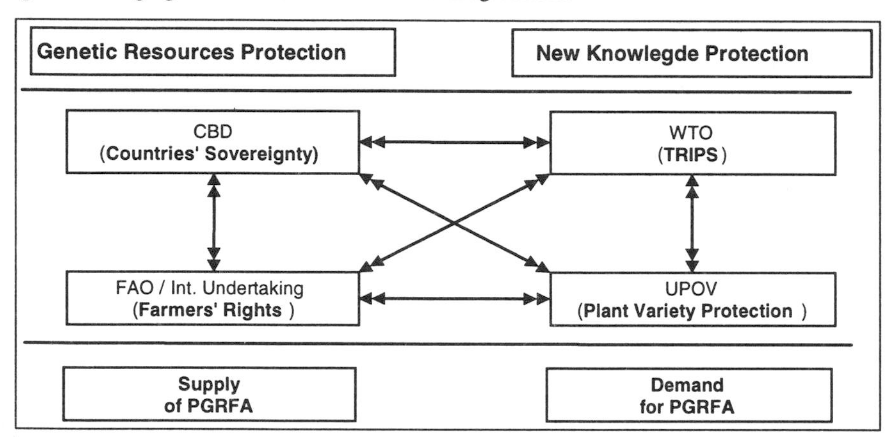

Another potential conflict arises between UPOV and TRIPS for countries which are members in both institutions. A double protection for different varieties of one crop species may occur because of different protection systems of different varieties of one crop, conflicting the breeder and farmer with respect to why the breeder's exemption and farmer's privilege exist for one variety but not for the other[40].

The TRIPS Agreement sets minimum standards for the crop plant protection for all member countries of the World Trade Organization. Under TRIPS, every member country must evolve a protection system, which, however, can be adjusted to a specific situation in a specific country. In contrast to UPOV, TRIPS is not an institution solely aimed at the breeders and representing their interests of protecting newly bred varieties. Hence, the sui generis legislation is the protection system, which might incorporate the compensation idea from the concept of Farmers' Rights and enabling the partially realization of benefit sharing (Leskien and Flitner, 1997). For instance, India is developing a sui generis system, aiming to incorporate the concept of Farmers' Rights into the system by defining the right for compensation for past contributions to conservation (Swaminathan, 1996b).

---

[40] The new concept of 'essentially derived variety' is the first attempt to solve a problem arising through technological change. Essentially derived varieties are varieties with single gene changes introduced by backcrossing or genetic transformation as defined by the 1991 UPOV Convention (Semon, 1995). The breeder's exemption is coming to its limit if modern biotechnology is utilized in plant breeding, provoking an inequality in competition. If a breeder inserts a patented gene into a protected variety, he may protect, and exploit commercially the modified variety, whereas if a breeder inserts a foreign patented gene into his own variety, he either has to pay royalties to the owner of the patent or could be prevented from exploiting the modified variety (Harries, 1996). Due to the concept of 'essentially derived variety' both breeders must seek to reach agreement with the other involved breeder.

It seems, however, unlikely that existing plant protection can be successfully used to define and enforce rights over traditional varieties because of the high variability and segregation of landraces which point towards a fundamental difference from protected varieties under the distinct, uniform, and stable criteria of UPOV (Lesser, 1994). Additionally, patenting landraces is not possible, because a patent must meet the criteria of novelty, which is difficult to certify for these genetic resources. A more feasible way is the definition of remuneration rights, with which not an exclusiveness is achieved but a compensation for contributions made by communities (Correa, 1994). Correa quotes an example of the blank tape royalty in some countries which applies to all tapes in order to compensate copies without the authors' agreement, while knowing that controlling private copying is impossible.

On the other hand, TRIPS' intellectual property system may interfere with the objectives of CBD through the possibility of patents for products resolved from genetic resources.

After the adoption of CBD, there was a need to harmonize the International Undertaking on Plant Genetic Resources with CBD. The main points of conflict are the realization of Farmers' Rights and the issue of access on mutually agreed terms to plant genetic resources, including the ex situ collections not addressed by the Convention (FAO, 1996d). The International Undertaking on Plant Genetic Resources has to be revised in harmony with CBD, because CBD is an international binding convention, whereas IU is presently a non-binding agreement, which will probably be attached as protocol to CBD after IU's revision (Esquinas-Alcázar, 1996).

The revised International Undertaking will have to take a firmer position in respect to financial mechanisms (FAO, 1995c). In other words, the access to PGRFA will have to be linked directly or indirectly to some financial or technology transfer. Hence, a revised International Undertaking will not support the cost-free access to PGRFA, therewith conflicting with the demand of UPOV for free access to PGRFA.

This conflicting situation has two implications for developing countries. In a time of increasing globalization of markets, it will be difficult for a country to refrain from signing TRIPS for its overall economic development. As discussed above, TRIPS offers the best possible solution for a legal framework enabling the protection of new knowledge and genetic resources. Therefore, TRIPS offers the most suitable existing framework for countries not only interested in protecting and "selling" PGRFA as raw material but also utilizing its information for breeding. Therefore, it is foreseeable that most countries will join TRIPS in the next ten years, which means that all of them will have to develop some kind of protection system for property rights, including modern varieties. Each country will have the chance to protect their PGRFA in some sui generis system. While, however, the enforcement for variety protection can be implemented in some way, it will be very difficult for any kind of PGRFA protection. The main difficulty is, as mentioned above, the proof of its originality and the implementation of any kind of exclusiveness and compensation as well as the consequent enforcement of any rights.

After harmonizing the International Undertaking with CBD on the one side and incorporating UPOV's concern into TRIPS, the present conflicting situation will probably dissolve; especially if biotechnology advances and genetic resources can be valued and handled according the genetically coded information. With further process in biotechnology it might even be that the patent system has to be simplified, because of confusion in patent agreements for one product (Shands, 1996).

In the actual discussion a range of ongoing legal instruments have been introduced for the access to germplasm in its different processing stages and the possibilities of benefit sharing derived from their use. In addition to the intellectual property rights, such as patents and plant protection, international binding and non-binding agreements exist on the access to, use of and remuneration for PGRFA such as FAO's International Undertaking on Plant Genetic Resources. Furthermore, material transfer agreements and other contractual agreements between individual gene owners or countries and biotechnology companies or countries as gene users may contribute to bilateral and/or multilateral approaches to the conservation of PGRFA (Barton and Siebeck, 1994). Other, non-IPR rights over intangible property, such as trade secrets, cultural property rights, remuneration rights, appellations of origin and protection of expression of folklore, also have to be considered by aiming at the implementation of Farmers' Rights[41].

4.4 Different Systems for the Transaction of Genetic Resources

Before CBD came into force in 1993, the instruments for the institutional framework of PGRFA conservation management, germplasm exchange and utilization were developed in a rather ad hoc manner, based mainly on national and international codex for research work. The achievement of having identified or discovered a variety in farmers' fields with interesting traits was credited through publications and other kind of awards; but the farmers who bred and maintained that variety in their fields were seldom mentioned. Germplasm exchange was regulated according to other natural resources transfer in research, i.e., free to all bona fide users and based on "pro mutua communatione", the mutual exchange as it is practiced between botanical gardens as well (Hammer, 1995). Engaged in the conservation and utilization of PGRFA since its beginning, FAO developed some instruments which are now integrated into FAO's *Global System for the Conservation and Utilization of Plant Genetic Resources for Food and Agriculture*. This global system was the formal framework for the access and exchange of PGRFA since the adoption of the International Undertaking on Plant

[41] For a more detailed analysis, see: Correa, 1994 and FAO, 1995c.

Genetic Resources, based on the undertaking's basic concept of a multilateral system (see Chap. 4.3.1).

Since the enforcement of CBD in 1993, however, the international exchange system for PGRFA has undergone some drawbacks. Because of its revision the framework given by the International Undertaking is not existent at present. Nevertheless, CBD provides for access to be granted on "mutually agreed terms", which might be agreed upon bilaterally or multilaterally (UNEP, 1994a). The current multilateral exchange system is generally conducted without a formal agreement. Hence, the existing exchange system is characterized by its informality and uncertainty. Nevertheless, most countries state that PGRFA in national collections are still freely available to all bona fide users. But in some Eastern European countries, the recent privatization of agricultural research institutes has increased the uncertainty over the continuing free availability of their PGRFA (FAO, 1996a). Furthermore, there are signs that the access to PGRFA in countries of supply is starting to be restricted, e.g., the Chinese genebank is restricting the exchange of indigenous germplasm (FAO, 1996a). Some other countries might restrict the access by establishing bureaucratic obstacles (Hammer, 1995).

There is presently a wide spectrum of genetic resources exchange systems in practice: on one side there is the informal multilateral approach, which does not reflect all of the legal binding points addressed in CBD; at the other end a strictly bilateral exchange system can be identified. Meanwhile partial elements of markets for genetic resources are emerging, as can be seen in contractual agreements between gene owner countries and gene users (either countries or the pharmaceutical industry in industrialized countries).

This leads to questions as to which arrangements for marketable components of genetically coded information might be practicable and how to create incentives for securing genetic resources which are not yet marketable.

Including the issue of the utilization's benefit sharing as well as the conservation's burden sharing, the future exchange system of PGRFA will determine the incentives for PGRFA conservation to a major extent. Only if the countries, hosting predominant parts of PGRFA, can anticipate a significant compensation for the costs of conservation, they will be prepared to invest in PGRFA conservation in the future as well. The countries' expectation will be determined to a major extent by the exchange system of PGRFA.

4.4.1 Multilateral System of PGRFA Exchange

The exchange of PGRFA presently takes place in a complex multilateral system which is based on legally non-binding and informal agreements, some of which were only accepted by tradition. The core element of the existing system is the unrestricted and free availability of PGRFA to all bona fide users worldwide. This policy, which is followed by all major ex situ collection facilities, was based on the general assumption that genetic resources were a *"common heritage of mankind"* and a *"common concern of humankind"* (UNEP, 1994, Preamble) and consequently a public good. Because of the free availability and determined by the concern of saving all threatened genetic resources, the conservation efforts were carried out in a very unsystematic way, determined mainly by the individual commitment of those involved. Furthermore, in the present multilateral system most countries share part of a total genepool of interest concerning most staple food crops and the governments have exercised little control over the exchange of germplasm in this informal system (Gass, 1996).

FAO and the IARCs combined with the NARS tried to systemize and link all conservation efforts on an international and a national level. Consequently, the main coordinators of the existing system in its present state are the "Global System for the Conservation and Utilization of PGRFA" and the network of genetic resource units and plant breeding activities, promoted and financial supported by the IARCs and the relevant NARS.

The global system consists of three pillars, based on the International Undertaking as framework agreement, and is accountable to the intergovernmental forum of the Commission of Genetic Resources for Food and Agriculture (FAO, 1997b):

- global mechanisms: an international network of ex situ collections under the auspices of FAO, a network of in situ and on-farm areas, crop related networks, and the World Information and Early Warning System on Plant Genetic Resources (WIEWS);
- global instruments: the international fund on Plant Genetic Resources for the implementation of Farmers' Rights, a periodic report on the State of World's Plant Genetic Resources, and a rolling Global Plan of Action on PGRFA;
- international agreements: an international code of conduct for plant germplasm collecting and transfer, a draft code of conduct for biotechnology, technical guidelines for the safe movement of Plant Germplasm, as well as the genebank standards.

The achievements of this system whose legal framework is still evolving, because of the obligation of renegotiating the International Undertaking, are the following:

- most of the earth's important PGRFA have been collected and conserved ex situ under this evolving, informal system (WIEWS, 1996);
- genetic resources have been utilized for breeding and provided the base for a tremendous food production increase (Wright, 1996);
- the system provided the fora for international negotiations on policies and regulations on PGRFA and enabled the emergence of the first report on the state of the world's PGRFA as well as the first global plan of action for the conservation and utilization of PGRFA;
- the first international PGRFA information system was created (WIEWS) and crop-specific and regional PGRFA networks emerged for cooperation reasons;
- the present exchange system with its free availability of PGRFA to bona fide users upon request enabled the advantageous utilization of more PGRFA from international sources than contributed to other national and international users because of the interdependence of all countries as regards the utilization of PGRFA.

Although some major points of conservation and utilization of PGRFA have been achieved with the existing system, there are still some major elements missing, those being the acceptance and implementation of specific financing mechanisms for benefit and burden sharing and the adoption of a revised International Undertaking including the recognition of the Farmers' Rights.

A multilateral system for controlling the exchange of PGRFA should have the task of providing suitable incentives for in situ conservation and ensuring a fair share of the benefits of the utilization of these resources to the donor countries. It should also facilitate access to plant genetic resources necessary for agricultural products and provide mechanisms for the regulation of this access where necessary, e.g., when requirements of nature conservation take precedence over other concerns. Such a system must also answer to basic institutional principles such as the equal access for the national agricultural research systems of all countries, the transparency of decision-making processes, the control of executive organs, and the securing of finances (Cooper et al., 1994).

The main advantage of a multilateral system of PGRFA exchange, in which all countries participate, is that it can only produce winners because each single participant gains access to more genetic material than he himself can contribute. Furthermore, a formal multilateral system of exchange may have an advantage over the present informal system, in that more transparency and certainty may exist with respect to the rights and obligations of the members on ownership, access and benefit-sharing (IPGRI, 1996). The present informal multilateral system is based on free access on the one hand and some – still not defined – compensation in the form of financial or a technology transfer or some kind of internalization on the other hand. The proposed multilateral system for exchange (MUSE) is conceptualized as a *"... framework for a system guided by a set of mutually agreed rules ..."* including bilateral and multilateral agreements (IPGRI, 1996, p. 61).

Not all participants, however, will profit equally from such a system. To base compensation merely on the number of species, varieties, or genes contributed to the multilateral system's common, internationally accessible pool would be too simple a solution. It would bear the threat of an adverse selection. Not all varieties will have the same valuable level of information. Some varieties will have more unique information than others. Hence, varieties with important, i.e., higher valued genetically coded information will be underpriced and will subsidize the less valued varieties, if all traditional varieties are treated as having the same value in a multilateral exchange system. If the countries or other suppliers of PGRFA are able to obtain the differences, they will try to take the more valuable varieties out of the multilateral exchange market and try to sell them independently on the basis of bilateral agreements. Consequently, the average quality of the regular PGRFA market will be reduced, which would lead to a reduction in the price. This tendency is increased by the proposed MUSE, as a framework in which different kinds of agreements can be signed, including bilateral contracts. Hence, the incentive will cause the countries to negotiate the most valuable germplasm in bilateral agreements because of the high demand, and the less demanded germplasm in the multilateral compensation pool.

The value of genetically coded information can never be determined a priori but rather only from an a posteriori observation, i.e., as a result of their success on the market. In order to remain viable the system must therefore sooner or later provide mechanisms of profit sharing.

4.4.2
Bilateral System of PGRFA Exchange

A bilateral agreement for the exchange of PGRFA is negotiated between two countries and is formalized through a contract. It can be restricted to a single exchange of germplasm between two countries, e.g., the bilateral agreement between Brazil and Malaysia for the exchange of a specific quantity of wild material of Hevea for a specific quantity of elite clones for other Hevea varieties (IPGRI, 1996). The bilateral agreement may also be a contract for a longer period of time, e.g., the Merck-InBio agreement on the collection, screening, and utilization of genetic resources from the Costa Rican tropical forest. The third potential category of a bilateral exchange system for PGRFA is the exchange of germplasm for financial resources. Hence, in addition to the time horizon, the exchange system may also differ with respect to the means of exchange: wild material for improved germplasm, wild material for technology or financial transfer.

According to IPGRI (1996) the main advantages of the bilateral exchange system are its flexibility in terms of negotiations, adoption and fulfillment of the contract, the good exploitation of respective comparative advantages, and low overhead costs because of missing permanent institutional structures.

The main structural disadvantage of bilateral agreements is related to the issue of benefit sharing. In contrast to pharmaceuticals, modern varieties are derived from other very different varieties, lines and germplasm. If benefit sharing must be realized by crediting the cascade effect involved in breeding, the benefit distribution through specific agreements seems to be operational, only if countries are involved as recipients from the benefits. But even then, there needs to be some intelligent infrastractural solution. Taking the Veery wheat released by CIMMYT in 1977 as an example, the complexity of a benefit sharing system may be visualized. The Veery wheat lines were developed from approximately 3170 crosses, made between 51 individual parents originating in 26 countries around the world; 62 varieties were released from the Veery lines and cultivated on approximately 3 million hectares around the world. (Skovmand, quoted in IPGRI, 1996, p. 26). Another question is whether every genetically coded information, integrated in the new variety at one time or the other, is equally credited or whether there should be any gradation depending on the novelty or the specific impact of the genetically coded information, i.e., will there be a possibility to assign a marginal value to each genetically coded information's contribution to a newly bred variety?

One of the most important arguments given by IPGRI (1996) and Cooper et al. (1994) for a multilateral exchange system is the problem of benefit agreements between a breeding company or a country and all the countries from which the incorporated individual parents originate, because of the high grade of interdependence of the countries in regard to the origin of the utilized PGRFA in a country. According to their arguments, the transaction costs would be too high for such an agreement. As mentioned above, through an intelligent agreement system the transaction costs may be reduced to a reasonable level. Furthermore, by adjusting the benefit agreements of PGRFA utilization to the breeders' protection rights, a practical solution may be feasible: if the Breeder's Privilege is kept in place, the benefits of utilizing germplasm of landraces should be restricted to new incorporated germplasm solely. Only those gene owners who supplied new PGRFA to the breeding success should be credited for their PGRFA and not those owners who have supplied PGRFA in the former breeding process while already incorporating their PGRFA before the release of an older variety. If the Breeder's Privilege is abolished and patents are introduced for all varieties, meaning that breeders cannot utilize an older variety for further research and development without crediting the owner of the incorporated old variety, the benefit system for genetically coded information derived from landraces has to be adjusted, i.e., all gene owners of germplasm derived from landraces through the whole breeding process have to be credited as well.

Additionally, an institutional framework is needed for a pure bilateral exchange system as well. The relevant information could not be distributed evenly without a clearing-house mechanism or some other system of information distribution. Hence, the transaction costs of the system would increase because of disproportionate search costs.

4.4.3 Market System of PGRFA Exchange

A market system for PGRFA will be a highly developed organization. This system must include the private interest in exchange of PGRFA and the social interest in the long-term conservation of PGRFA. There will not only be a need for an appropriate institutional framework, but the system will also require a practical system of enforcement and control.

The main institutional aspect for a private market system is the creation of a virtual market for genetically coded information on the internet as a freely accessible database containing the World's entire available PGRFA (e.g., the World Information System and Early Warning System on PGRFA). Prospective users could call up the standardized genetically coded information they require and clarify the property rights of the resources. Prices for genetically coded information would develop stepwise depending on the technical advances made by users and the development of markets for end products, amongst other factors.

In addition to the information system a further system is needed to enable the transaction of genetically coded information from the supply to the demand side as well as to enforce any kind of compensation. This system could be a clearing house mechanism for the promotion of technical and scientific cooperation, serving the sustainable development of biological resources as mentioned in Article 18 of CBD or a "facilitator" to organize the transfer of genetically coded information, as was suggested by Lesser and Kraittiger (1993). One successful example is the "International Service for the Acquisition of Agri-Biotech Applications" (ISAAA), which as an institution is playing the role of "an honest broker" between the different interest groups in technology transfers (van Zanten, 1996).

4.4.4 Determinants for the Development of Different Exchange Systems and Their Funding

Whereas the pharmaceutical industry competition for access to genetic resources is leading to the negotiation of bilateral contracts between large corporations and national governments, the agricultural sector appears to be moving towards a further consolidation of the existing multilateral system because of its structure and the high transaction costs entailed therein (Cooper et al., 1994).

The bilateral and the private exchange system of PGRFA may emerge from this system in the long run, after further development in the patent law and in property rights in general as well as in technology for the enforcement of any rights[42]. Furthermore, it will eventually be difficult to establish an exchange system recognized overall because of the heterogeneity of developing countries. The uneven distribution of agrobiodiversity among developing countries and the different grade of technology development discloses the fundamental controversy over internalization or compensation for the access to PGRFA.

Each country will favor an exchange system depending on its stock on PGRFA as well as on its ability to utilize their ex and in situ maintained plant genetic resources. The national capacity of genetic resources utilization depends on the country's technological state of crop improvement programs at present and in the future. Knowledge, infrastructure, and financial potential for meeting the requirements of further breeding activities, are the main determinants, which differ from country to country. Consequently, the prospects for the technological capacity for utilization of plant genetic resources in countries are diverse.

According to their breeding capacity and the availability of genetic resources the national utilization potential differs greatly from one country to another, which suggests a classification into 8 groups of countries. Countries can be roughly grouped into plant genetic resources rich and poor countries[43], as well as into 4 categories corresponding to the different levels of development of crop

[42] Basically, the techniques are now available to utilize molecular markers in the context of variety protection (van Laecke et al., 1995). Further development of molecular techniques will theoretically improve the level of distinction and allow researchers to distinguish even between closely related cultivars. This may allow the identification of essentially derived varieties (Semon, 1995). Hence, modern techniques may prove to be useful in determining the probable identity and origin of landraces, genotypes and genes. However, it is unlikely that these techniques can be utilized in the context of agreements for access to PGRFA due to the inherent great variability in most landraces and populations (FAO, 1995).

[43] See Chap. 3.1 for the concept of agrobiodiversity-rich and poor countries.

improvement programs based on the information provided in the Country Reports (see Table 4.1)[44].

Seven % of all reporting countries (152) do not have sufficient resources to develop crop improvement programs. These countries are classified as countries with *no programs*. Smaller countries of sub-Saharan Africa or the Small Island Developing States of the Pacific are examples of those countries (other examples can be seen in Table 4.1), which are not able to utilize their own existing plant genetic resources potential. These countries will depend further on the finished products, developed in other countries, even though they may have a potential of plant genetic resources, but the low demand for modern seed and limited resources are not enough incentives for the development of independent programs.

Basic crop improvement programs, including germplasm identification, introduction and evaluation programs, are predominantly performed in the public sector of countries, where farmers rely mainly on vegetatively propagated crops, e.g., African and Caribbean countries. This group is made up of 26 countries (17% of the reporting countries).

Countries in Eastern Europe, Near East, Asia, Africa, and the smaller countries of Central and South America, have the capacity to carry out *well developed crop improvement programs*, i.e., evaluation, enhancement and improvement of PGRFA through hybridization programs. This group is made up of 74 countries, where the activities are still publicly dominated (49% of the reporting countries).

Countries from East Asia, Latin America and a few other countries such as India, the Philippines, Kenya, and South Africa are to be found in the category of

Table 4.1. Classification of developing countries according to their breeding capacity and their PGRFA potential, examples

	Crop improvement program status			
	No programs	Basic	Developed	Advanced
Genetic resources poor countries	Central African Republic	Angola, Namibia, Cape Verde	Burkina Faso, Uruguay,	
Genetic resources rich countries	Rwanda	Myanmar, Papua New Guinea	Bolivia, Madagascar, Ethiopia, Korea	Venezuela, Brazil, India,

See Chap. 3.1 for the concept of agrobiodiversity-rich and poor countries.
Source: FAO, 1996a and all relevant Country Reports

industrialized countries *advanced crop improvement programs*. A good level of technological and scientific competence exist in these countries, including the

[44] The 4 categories corresponding to the different levels of development of crop improvement programs are based on the information provided in the Country Reports and FAO, 1996a.

access to plant biotechnology (this group is made up of 41 countries or 27% of reporting countries). Especially developing countries with more financial resources and personnel trained like China, India, the Republic of Korea, Argentina, Brazil and Mexico are increasing their technological capacity (von Braun and Virchow, 1997).

Consequently, developing countries rich in PGRFA and with developed or advanced crop improvement programs presently have comparative advantages for an internalization of the benefits resulting from PGRFA utilization and could win through a bilateral or private exchange system. Developing countries, rich in PGRFA but poor in technology potential, could win through the internalization of benefits, if adequate enforcement systems are implemented. On the other hand, because of high transaction costs for such an enforcement, these countries could also win through the compensation solution in a very broad multi-lateral system. PGRFA-poor developing countries, however, will push a compensation solution, based on a multi-lateral system in which compensation is allocated by unspecified criteria. This would be the only means which those countries could benefit from the international exchange of PGRFA, besides benefiting from the supply of modern varieties through international public and private channels.

Nevertheless, regardless of which exchange system will dominate, key requirements for any system are (1) the secure conservation of PGRFA (in situ as well as ex situ) on a national and an international level, (2) an ensured system for crediting the supplied PGRFA, and foremost (3) a comprehensive information system, which enables the reduction of transaction costs for PGRFA exchange as well as the enforcement of any benefit agreements.

The system of internalization would tend to a bilateral exchange system or imply a market approach, where either the countries or the farmers, individually or as community, will profit directly from any germplasm exchange. Meanwhile the system of compensation would suggest a more general, non-targeted approach for a multilateral exchange system. One key role for the compensation system is the establishment and utilization of an international fund as the nodal point for financial transfer from PGRFA users (countries or private companies) to PGRFA suppliers. The basic framework for the "International Fund for Plant Genetic Resources" was adopted by FAO Resolution 3/91, which agreed *"... that Farmers' Rights will be implemented through an international fund of plant genetic resources, which will support plant genetic conservation and utilization ..."* (FAO, 1996f, paragraph 16). Matters related to the legal status, policies, priorities, and parties as well as to the financing are, however, still under discussion. In the discussion, it is of importance to differentiate between questions concerning the sources of funding on the one hand and the mechanisms of imposing and distributing the funds on the other.

The majority of industrialized countries would like to see the funds already allocated to multi- and bilateral development assistance as the major part of the benefit sharing in a multi-lateral exchange system. For this reason, the industrialized countries have difficulties in accepting a new fund; they prefer rather the funding organized and directed through the existing channels. Among these are the bilateral official development assistance, multilateral and international development assistance through the UN organizations and their funds, mainly FAO, UNDP, and UNEP, regional and other development banks and their funds, e.g., the World Bank, GEF, and IFAD. NGOs, foundations, universities, and research institutes contribute to the conservation and utilization of PGRFA as well[45]. The developing countries, however, call for new additional funding for the conservation of PGRFA. They do not accept the present funding as part of the claimed benefit-sharing in CBD and from the further revised International Undertaking[46].

New and additional funding may be derived from very different sources. Indirect financing mechanisms could be realized through programs such as debt reduction and special bonds, investment rights, and joint implementation programs. On the other hand, direct financing of the international fund could be made feasible by raising funds through international levies or charges with relation to specific markets. This is legally possible through the implementation of what are called "Remuneration Rights" (Correa, 1994). For instance, taxes imposed on the profit or turnover of the private (and public) seed sector or on agricultural products could be sources of financing for the international fund. Swaminathan (1996c) suggests that technology-rich countries should add 0.01% of their GDP to their official development assistance for a "Global Fund for Biodiversity Conservation for Sustainable Food Security" or a 2% levy on all seed sales (Swaminathan, 1996b). Furthermore, a 1% assessment on all agriculture-related products should be transferred to "National Community Gene Funds" for the realization of Farmers' Rights (Swaminathan1996c).

As an additional source of funding, FAO (1996b) suggests an increase in the efficiency of existing funding, e.g., to reduce duplication of efforts and a reallocation of existing bilateral and multilateral financing and domestic agricultural expenditures.

A further point yet to be solved is the mechanism and criteria for the fund distribution. The mechanism for the distribution, through existing or new channels, is of only marginal importance at present. The criteria of distribution are of major relevance. FAO (1996b) proposed the criteria and priorities for funding as agreed to by the first Conference of the Parties to the Convention on Biological Diversity (UNEP, 1994b). These criteria are still, however, very general; a priority can, therefore, not be identified in following these criteria.

Regardless of how the international exchange system and financing mechanisms for the management and utilization of agrobiodiversity are ultimately framed, they will always require a concept according to which an external monitoring and evaluation system by which instruments could be constantly and purposefully evaluated and controlled. Without such a mechanism there is a high risk that the financial resources provided are spent inefficiently or consumed by high transaction costs.

---

[45] See Chap. 5.3 for more detailed information.

[46] These differences were the main reason for almost breaking off all the discussions at the International Technical Conference on Plant Genetic Resources in Leipzig.

In this chapter, the institutional framework between supply and demand and its development has been analyzed. The lack of or poorly defined property rights for genetic resources is the major institutional factor determining the threat of PGRFA diversity loss. PGRFA's future conservation and exchange will depend, however, on the sufficient incentives as a kind of compensation or internalization. The development of the institutional framework is promoted by different legal agreements and accompanied by different options for an exchange system. The debate which incentive system and exchange system to choose is the debate between agrobiodiversity-rich and diversity-poor countries as well as between technologically rich and poor countries. It seems to be the first step towards a market or bilateral exchange system, however, the existing informal multi-lateral exchange system will be the most likely system to be evolved into a formal exchange system.

5 Conservation Costs of PGRFA

The overall costs of PGRFA conservation are made up of the fiscal costs and the opportunity costs (see Fig. 5.1). The fiscal costs, representing the costs arising for PGRFA conservation which have to be budgeted and invested either on national or international level, for planning, implementing and running ex situ and in situ conservation activities, are determined by the specific conservation activities, the depreciation costs for investments, and the costs for institutional and political regulations for access to PGRFA. Additionally, costs for compensation and incentives paid for maintaining PGRFA have to be reflected. Furthermore, there are the opportunity costs, reflecting the foregone benefit for the country by maintaining the diversity of genetic resources in the field.

Fig. 5.1. Economic concept of the costs of PGRFA conservation

Costs of PGRFA Conservation
Opportunity Costs
Fiscal Costs
Opportunity costs for reduced production
Costs for compensation / Incentive
Conservation expenditures
Depriciation costs
Costs for inst. & political regulations
Costs relating to in situ activities
Costs relating to ex situ activities

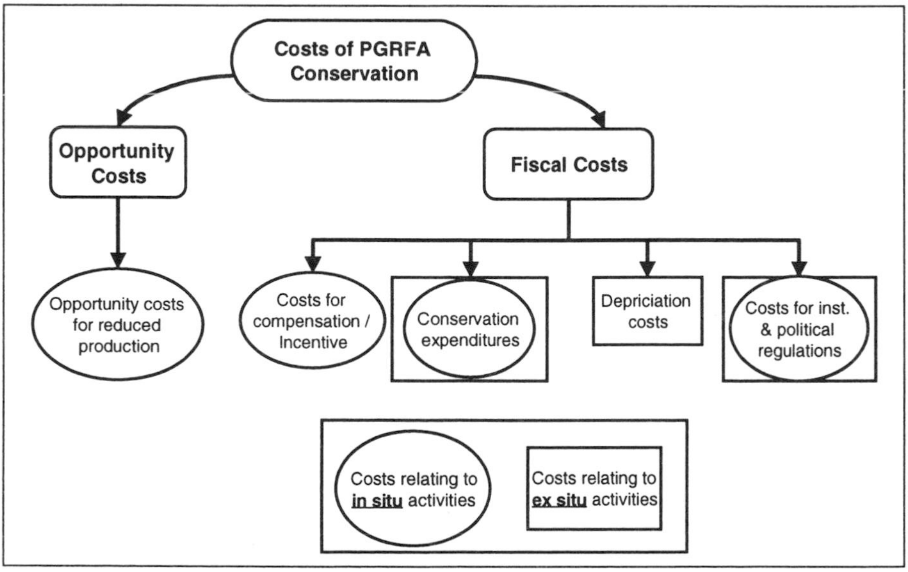

It is important to reflect both elements of costs for the discussion of a cost efficient conservation of PGRFA. A national program for PGRFA conservation may have low fiscal costs due to prioritizing a low-cost in situ conservation method, i.e., by withholding agricultural development from farmers in marginal areas. But including the opportunity costs of the foregone food production in the overall cost calculation, a different national program, based on a more cost-intensive ex situ conservation system, may have less overall conservation costs due to reduced opportunity costs. Consequently, the national and the global welfare could be better off with that system.

Different approaches may be taken in identifying the specific costs for the conservation of PGRFA. Costs can be identified on different levels as well in different categories. Considering the players in the conservation activities, costs can arise at the farm level or at national and international levels as well as at the level of conservation activities in the private sector. Conservation costs may, however, occur in different categories, depending on the conservation methods used: costs of in situ and ex situ conservation as well as costs for the supporting activities and institutional process for the conservation of and access to PGRFA. Therefore the method of estimating costs depends upon the approach taken. In view of the inherent difficulties and limitations in compiling data on the current expenditures on PGRFA, the purpose of these costing estimates were confined to obtain orders of magnitude of the current efforts made. Hence, no attempt was made at this stage to differentiate one-time expenditures in the form of investments or fixed-term projects for capacity building from running expenditures for PGRFA programs (e.g., gene bank maintenance). In Chap. 5.1, the costs of conservation will be identified for particular cases according to the different cost categories. The costs on national level will then be presented for specific countries - supplemented by a rough estimation for the overall costs of all countries involved in PGRFA conservation (Chap. 5.2). In concluding on the cost aspect, the incremental costs on international level are pointed out in Chap. 5.3. An analysis of the effectiveness of conservation in specific countries will be combined with the cost analysis of the countries involved and the results are discussed in Chap. 5.4.

5.1 Analysis of the Costs for PGRFA Conservation According to Different Categories

Conservation costs arise from the different ex situ and in situ conservation activities as well as all supporting activities including the institutional process determining the access to PGRFA and the cost and benefit sharing of the conservation and utilization of PGRFA. Hence, the following cost calculations will be presented for particular cases of conservation activities. By differentiating the costs according to the place of origin, all conservation costs involved can be identified (see Table 5.1). The costs involved are compiled according to three main categories: labor costs - predominantly for the ex situ conservation; costs for area demand - partly as opportunity costs and partly as compensation payments to farmers for in situ conservation; costs for conservation material and investments - mainly for ex situ conservation.

Table 5.1. Costs arising according to different categories

Ex situ:	In situ:
Acquisition costs	Direct costs for in situ conservation programs and projects:
Maintenance costs	Production costs
Processing costs	Costs for supporting activities, incl. for compensation payments
Administration costs	Costs for implementation and management of in situ programs and projects
Opportunity costs:	
Supporting activities and regulations:	
Costs for institutional and political negotiations	
Costs for implementation and maintaining a PGRFA exchange system	
Costs for information and control management	
Harmonization of national policies	

5.1.1 Costs of ex Situ Conservation

Costs for ex situ conservation activities depend upon the specific crop to be conserved and the specific conservation method applied, as discussed in Chapter 2.4. The ex situ conservation costs range widely depending on conservation site, material to be conserved, the handling required, and specialized facilities. For instance, it is more difficult and therefore more cost-intensive to maintain the genetic integrity of cross pollinated crops than self-pollinated crops during regeneration (Porceddu and Jenkins, 1991). The costs are mainly determined by the salaries, which vary regionally depending on labor costs. Other costs differ as well, determined by regional price structures, e.g., the capital costs for basic and working equipment (including buildings and the area needed for regeneration etc.), the running costs - particularly energy for cooling or freezing, durable material, and the maintenance of the equipment - and the costs for administration activities.

Some cost estimates do exist, of which each reflects a very particular case, differing from the others because of the incorporation of only specific activities of the above mentioned four broad cost components and because of the different crops conserved and sites of conservation. The most significant estimates are presented and discussed in the following section. The marginal costs, determining the potential supply pattern, are the most important costs for an economic analysis. The unit costs of one accession seem more telling for future investment decisions taking the following three factors into consideration: (1) the absence of a market for PGRFA; (2) the limitation of available storage capacity in the major genebanks; and (3) the actual need for new storage facilities or the extension of existing facilities.

Furthermore, average total costs per accession seem the more adequate approach in calculating national and international ex situ and in situ conservation costs and aiming to obtain orders of magnitude of the current annual efforts.

Glachant (1991) published one of the first general cost estimates for ex situ conservation of seeds (see Table 5.2). The costs are differentiated according to the capacity of the seedbank. Due to economies of scale in the storage activities, the conservation costs are the lowest in the biggest storage facilities, capable of conserving over 50,000 accessions. The costs were estimated for national and international genebanks, but included only some of the major tasks of conservation activities. The estimate excludes pre-conservation activities (most of the above defined acquisition costs) as well as the costs of processing and administration. Therefore, the evaluation is only partially included. Consequently, Glachant's estimate of US $ 16 per accession and year can be interpreted as pure conservation costs for orthodox seeds with limited additional information.

Table 5.2. Ex situ conservation costs according to seedbank capacity

Seedbank capacity in accessions:	Costs /accession in seedbanks (US $ / a):
< 10,000	22
10,000 – 50,000	23
> 50,000	6
Average of all seedbanks:	16

Source: Glachant, 1991

A much more precise cost calculation was undertaken by Trommeter (1993) (see Table 5.3). Besides the plain conservation costs, he includes the acquisition costs and some of the processing costs. Unfortunately, Trommeter gathers cost information from different regions - the conservation costs for the fieldbank originate from the Ivory Cost, all acquisition costs evolve from activities in Central Africa, the other storage facilities originate in France - and from different crops (millet and sorghum for the seedbank, coffee for the fieldbank). Even though the crops conserved differ from each other, one may regard them as typical samples for each storage facility. Costing for ex situ conservation has been carried out for potatoes in Germany (Gehl, 1997a). Potatoes are one of the important crops worldwide for food energy supply, with high regional importance, e.g., Europe, South-America (FAO, 1991). Large collections of potato accessions are to be found in Colombia and the USA, but 20% of the world collection of 31,000 accessions is held by CIP under medium-term in vitro storage (6257 accessions) (FAO, 1996a). Besides CIP and Colombia, Germany has one of the largest potato collections (13%), stored in Gatersleben at the Institut für Pflanzengenetik und Kulturpflanzenforschung (IPK) and in Braunschweig at the Bundesanstalt für Züchtungsforschung (BAZ).

Table 5.3. Ex situ conservation costs according to different storage facilities

Kind of genebank:	Costs/accession (US $ / a):
Seedbanks[a]	80
Fieldbanks[b]	45
In vitro[a]	274

Costs originate from: [a] France; [b] the Ivory Coast[47]
Source: Trommeter, 1993

Gehl has been costing the different ex situ conservation methods at BAZ and at IPK (see Table 5.4). Similarly to Glachant (1991), Gehl concentrates on pure conservation activities calculating the maintaining and administration costs, while neglecting the acquisition and processing costs. Because of the quality of cryopreservation and its high safety, the cost-intensive conservation activities, i.e., the preparation for freezing, which are responsible for over 90% of the total costs, can be reduced to once every 50 years or even longer (Schäfer-Menuhr, 1996). Hence, these preparation costs may be distributed over the whole conservation period of 50 years as annuity. Consequently, cryopreservation shows the lowest annual unit costs for ex situ conservation with US $ 22 per accession and year.

The majority of the cost accounting presented for ex situ conservation facilities were made in industrialized countries (France and Germany). Therefore the conservation costs cannot be simply applied for calculating the overall conservation costs for ex situ storage facilities worldwide. For a first rough approach, the countries conserving PGRFA ex situ are divided into two groups: the group of

Table 5.4. Costs for ex situ conservation of potatoes in Germany

Conservation Method	Costs in US $/ accession / year		
	Variable costs	Fixed costs	Total costs
Cryopreservation[a]	0.04	22,26	22.29
Field-genebank[b]	11.18	39.41	50.59
In-vitro / slow growth	10.00	127.65	137.65

[a] recalculated according to information from GEHL, 1997b (see Appendix 7)
[b] managed by IPK, Gatersleben, all other activities managed by BAZ, Braunschweig
Source: Gehl, 1997a

[47] Trommeter originally calculated US $ 136 for the conservation of one accession in a field genebank (Trommeter, 1993, p. 60). The pre-conservation costs were excluded to make the figures comparable.

industrialized countries – such as the OECD group - and the developing countries. Even though the OECD countries hold a large share of the world's total ex situ conserved PGRFA (see Table 5.5)[48], the existing conservation capacity of developing countries and their objective to increase their share of conservation has to be taken into consideration as well[49]. Hence, the cost estimate for all ex situ conserved accessions has to be based partially on costs, which reflect the cost structure of developing countries.

Table 5.5. PGRFA conserved ex situ in OECD countries

OECD countries in region	Number of accessions	Percentage of accessions held by OECD countries in regional category
Total OECD in Europe:	1,021,641	53%
Total OECD in America:	836,148	60%
Total OECD in Asia / Pacific:	202,581	13%
Total OECD countries:	2,060,370	
Worldwide stored accessions (without CGIAR's Centers)	5,554,505	37%
Wordwide stored accessions (with CGIAR'S Centers)	6,147,696	34%

Source: calculated according to information from FAO, 1996a, SGRP, 1996, WIEWS database, 1996[50]

In cooperation with some conservation experts in India, cost estimates were made for storage facilities in the Indian context. The costs are based on the conservation activities of ICRISAT. Data was added by conservation activities of the MS Swaminathan Foundation in Madras, India. The attempt was made to estimate the costs for each of the specific tasks in the conservation process. Following the cost structure for ex situ conservation outlined above, the costs may be summarized as follows (see Table 5.6):

[48] 37% of all 5,554,505 accessions worldwide are stored in one or the other OECD country. Even when accessions stored by the CGIAR Centres are included in the calculation, 34% of all 6,147,696 accessions are conserved by OECD countries.

[49] 72 countries indicated in their Country Reports their need for new or extended storage facilities (respective Country Reports; FAO, 1996a).

[50] For more detailed information see Appendix 5.

Table 5.6. Key parameters for ex situ conservation costs on genebank level

	Underlying assumptions	US $ / Accession	US $ / Acc. / year
Acquisition costs:			
Survey, inventory, collection, transportation[a]	For 100 years	100.00	1.00
Multiplication, information gathering[a]	For 100 years	0.50	0.01
Characterization and 1st Evaluation[a]	Once every 10 years	3.00	0.30
Maintenance costs:			
Conservation preparation, conservation, and germination control[b]			0.57
Regeneration[b]	Once every 10 years	3.50	0.35
Processing costs:			
Information management[c]		0.14	0.14
Systematically information record: 2nd Evaluation (per each stress trait)[a]	Average of 5 traits for each acc. every year	3.00	15.00
Multiplication (including packing and shipping)[a]	Once every year needed	0.35	0.35
Administration and and general costs:			
Maintenance of storage facility[c]		0.19	0.19
Capital depreciation[c]		0.15	0.15
Subtotal:			18.06
Administration overhead costs	10% of other costs		1.81
Total:			19.86

Source: Calculated according to information from: [a] Singh, 1996; [b] Rao, 1996; [c] Rani, 1996

1. **Acquisition costs**: The precondition of all conservation activities is to survey and categorize as well as to collect and transport existing PGRFA of a specific (agro-ecological) area to the genebank. Experience shows that a collection mission of 60 days may "harvest" between 120 and 150 varieties. The costs for such an international trip will exceed US $ 15,000 (Singh, 1996). That makes an average of US $ 100 for each accession. Because of the underlying concept of ex situ conservation, there should be no need for further collection of germplasm already collected. Therefore these costs can be distributed as annuity over the expected conservation period. If the first objective of PGRFA conservation is accepted, i.e., "freezing for future utilization" (see Chap. 2.4), then a conservation span of 100 years must be presumed, which reduces the annual

costs of germplasm collection to US $ 1.00 per accession. No discount rate is applied for this first rough estimation. At the genebank, the material collected is multiplied to increase the seed availability, to generate the two safety-duplicates, as well as to collect information on its characterization and the first evaluation. The multiplication costs, including the preparation activities, are calculated with US $ 0.5 for one accession (Singh, 1996). These activities occur only once in the conservation span of the accession. Therefore the costs can be distributed as annuity over 100 years, which amounts to costs of US $ 0.01 per annum. The costs for characterization and the first evaluation of the accessions amount to US $ 3.00 per accession (Singh, 1996). These activities must be repeated every time the accession is regenerated to verify the germinated material. It is a common practice to regenerate the germplasm every 10 years. Consequently, the annual average costs of characterization and first evaluation are US $ 0.3.

2. **Maintenance costs**: The seed for one accession to be stored in the genebank and two safety duplicates stored in other genebanks must be prepared for the conservation (dried to 5% moisture content, packed and sealed). Once stored in the genebank (for long-term storage) annual costs for electricity and germination control in certain frequencies are incurred. US $ 0.57 are calculated for the conservation preparations, the conservation, and the germination control (Rao, 1996)[51]. As a second important activity for maintaining germplasm, the accessions are regenerated once every 10 years. Theoretically, accessions may be stored without loosing their viability for up to 100 years or even more (Holden et al., 1984). It is difficult to generalize the cost for regeneration, because the regeneration activities are very site-specific. If the accessions are regenerated at a site at or nearby the genebank costs amount to US $ 3.5 for one accession (Rao, 1996). The more distant the place of regeneration and the more difficult the pollination of the crop, the higher the costs, which may exceed US $ 50 (CGRFA, 1996). For a cost calculation of a genebank located in a developing country, it is assumed that most of the germplasm originates from the country itself, resulting in minimal costs.
3. **Processing costs**: Besides the conservation activities in the narrow sense, costs are incurred for preparing the accessions to be easily accessible for utilization as well as for systematically recording information and management. According to Rani (1996) the information management may be calculated with US $ 0.14 per stored accession and year. Furthermore, US $ 3.00 are calculated for the analysis of each single trait in the 2nd evaluation; i.e., the screening for biotic or abiotic stress factors (e.g., certain salt tolerance) or for food qualities (e.g., for oil or protein) (Rao, 1996). It may be estimated that an accession is screened on average for up to five traits every year. Therefore, the annual calculation amounts to US $ 15.00 per accession. By request, the germplasm is multiplied and sent to the requesting organizations. Because of the evaluation process the

---

[51] According to RAO the amount is US $ 0.45 per accession and year, but this estimation includes only the electricity, the maintenance of building and equipment, and the personnel costs for the genebank manager. An additional US $ 0.12 per accession has to be calculated to cover all accession maintenance work.

germplasm also has to be multiplied every year. The multiplication aimed at obtaining information of the cultivated genetic resources is more work intensive than the pure multiplication for shipment. Multiplication, including packing and shipping, has been calculated at US $ 0.35 under the assumption that an accession is requested once every year (Singh, 1996).

4. **Administration and other costs:** In addition to the administration costs, which are calculated as overhead costs of 10% of the other costs, the storage facilities have to be maintained. US $ 0.19 per accession are estimated for the administration (Rani, 1996). Furthermore, the estimated average annual capital depreciation amounts to US $ 0.15 (Rani, 1996).

Following this approach, the annual overall costs for the ex situ conservation of an accession in a developing country's seed genebank are estimated to be US $ 19.86.

To estimate the overall conservation costs for ex situ stored PGRFA, it is necessary to differentiate the stored accessions according to the storage type (see 3).

Fig. 5.2. Storage type of ex situ conserved PGRFA (in %)

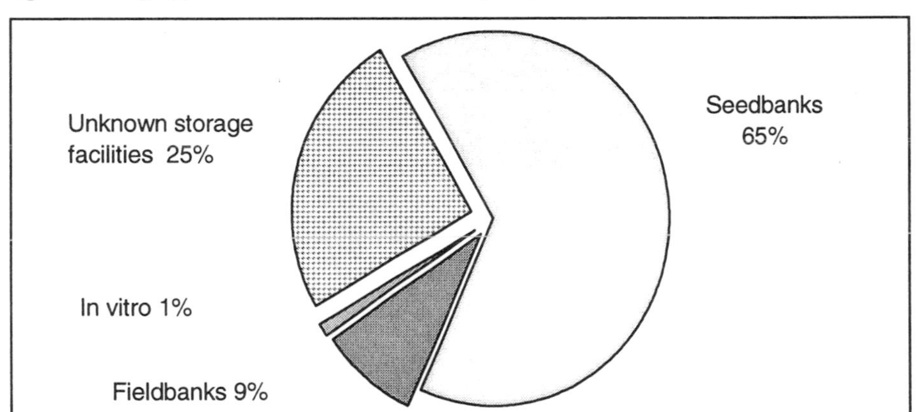

Source: calculations according to information from FAO, 1996a; FAO, 1995d; WIEWS, 1996[52]

Most of the 5,554,505 accessions are stored in seed genebanks - 3,610,428 accessions (WIEWS, 1996). Next to the seed genebank, the field genebanks are the most common ex situ conservation method particularly for vegetatively propagated crops which require regular replanting and whose long-term security is uncertain. Therefore, 526,300 accessions are stored in the more labor-intensive and costly field genebanks (FAO, 1996a). In vitro storage is developing into an alternative or complementary method for conserving vegetatively propagated crops as well as those with long lifecycles or recalcitrant seeds. Successful in vitro collections exist for plantain, banana, cassava, yam, potato, sweet potato, and *Allium* spp. Because the technology requires expensive equipment and skilled staff, and because it has

[52] See Appendix 6 for more detailed information.

only been developed for a relatively small number of species, there are only a few storage facilities utilizing in vitro conservation. Therefore, the amount of conserved accessions in vitro is still very small. Approximately 37,600 accessions are stored in this way (FAO, 1995d).

For approximately 1,400,000 ex situ conserved accessions (25%) the information concerning the storage type is missing. As can be seen from the Country Reports, most of the accessions without detailed storage type information may be assigned to orthodox seed and consequently may be assumed to be stored in some seed genebank facilities. It is assumed that approximately 90% of these accessions are stored in seed genebanks, 8% in field genebanks and 2% in in vitro storage facilities. Based on these assumptions, 88% of all worldwide conserved accessions are stored in seed genebanks and 11% in field genebanks, whereas only about 1% are stored in in vitro facilities and less than 0.2% of the accessions are conserved with the technique of cryopreservation[53] (WIEWS, 1996; FAO, 1996a).

With the rough knowledge of the distribution of accessions according to the different storage types, this information can be combined with the distribution of stored collections according to the grouping into OECD and developing countries (see Table 5.5). In this way, a rough cost estimate may be carried out, taking into consideration the different levels of costs. Hereby it is assumed that these differences may be captured by the two different levels of cost calculations made (Table 5.7).

By dividing the worldwide conserved accessions according to storage type and location and, simultaneously, assigning the different costs to the corresponding conservation technologies and locations, the total costs of all ex situ conserved PGRFA may be estimated. Nearly US $ 250 million are spent annually for the global ex situ conservation of agriculturally relevant germplasm. Approximately 85% of the costs are spent for seed genebanks, which store nearly 90% of all accessions. The average costs per accession were calculated for each storage type. An overall unit cost may be calculated as well. This unit cost amounting to US $ 44 per accession at the international level may be used as a calculation basis for obtaining an overview of the costs for conserving PGRFA ex situ.

---

[53] See Appendix 11 for more detailed information.

Table 5.7. The average costs for ex situ conserved accessions

	Kind of storage facility:				
	Seed genebank	Field genebank	In vitro	Cryopreservation	Total
General distribution of accessions according to storage type	88%	11%	1%	0.2%	100%
Number of accessions in:					
OECD countries	1,802,824	226,641	26,785	4,121	2,060,370
developing countries	3,057,368	384,355	45,424	6,988	3,494,135
Costs/accession (US $) in:					
OECD cost level	80.00 [a]	50.59 [b]	137.65 [b]	22.29 [e]	
Developing countries	19.86 [c]	45.42 [a]	82.92 [d]	18.23 [f]	
Total costs (US $) in:					
OECD countries	144,225,900	11,465,753	3,686,929	91,851	159,470,433
Developing countries	60,720,860	17,456,116	3,766,538	127,396	82,070,910
Global total costs (US $)	204,946,760	28,921,869	7,453,467	219,247	241,541,343
Global total cost distributed according to storage type	85%	12%	3%	0,1%	100%
Average costs per accession (US $)	42.17	47.34	103.22	19.74	43.49

[a] see Table 5.3; [b] see Table 5.4; [c] see Table 5.6; [d] see Appendix 7; [e] seeAppendix 8; [f] see Appendix 9

Source: calculated according to the above-mentioned sources

5.1.2 Costs of in Situ Conservation

The discussion on the in situ conservation of biodiversity is mainly centered on setting aside state or private land in parks, reserves and forests (Norton-Griffiths and Southey, 1994). Programs and projects on in situ conservation are classified into 5 categories: (1) on-farm conservation by farmers without any external incentives, (2) protected areas for conservation of wild relatives of crops, (3) projects with strong farmer participation, (4) conservation with emphasis on utilization, and (5) community-based genebanks (Cooper et al., 1994). It must be noted that the last category cannot be classified as in situ conservation, resulting in only 4 general categories. According to the information in the Country Reports, only 17 of all 145 surveyed countries have some kind of in situ conservation or the other (FAO, 1996a).

The costs involved in the different in situ conservation activities are:

1. the implementation and management costs for the programs and projects, e.g., establishment costs, running costs of infrastructure, materials and personnel;
2. the costs for compensation or incentives paid for maintaining PGRFA diversity in farmers' fields;
3. the opportunity costs for the sacrifice of land for the maintenance of farmers' varieties and consequently the loss of food production through the abandonment of high-yielding varieties.

Up to the present, cost estimates for in situ conservation are rare. An interesting new study on the costs of in situ conservation is the survey on potatoes in Peru by Gehl (1997). It was estimated that the conservation of one traditional variety costs US $ 594 per year (see Table 5.8); 84% of these costs are fixed, determined by the in situ conservation project and its costs. Only 16% of the costs (US $ 98) were derived from the opportunity costs, defined as foregone benefit by cultivating the traditional variety, supposing that farmers cultivated the different traditional varieties because of conservation concerns.

Table 5.8. In situ conservation costs for potatoes in Peru

Conservation Method	Costs / accession / a in US $		
	Variable costs	**Fixed costs**	**Total costs**
In situ conservation / Peru[a]	98	496	594

[a] Including opportunity costs
Source: Gehl, 1997a

Even if the opportunity costs are low compared with the fixed costs, it must be stressed that US $ 98 per variety maintained in situ is still more than 100% higher than the average costs for conserving one accession ex situ (US $ 44).

The on-farm conservation by farmers without any external incentives is the most important way of conserving PGRFA at present. For instance, the Arguarana Jivaro community in the Peruvian Amazon grows 61 distinct cultivars of cassava, while some small communities in the Andes grow 178 locally named potato varieties (Brush, 1991; Bolster, 1985). From the national perspective and if alternatives are created, the area cultivated with the traditional varieties should be minimized by maintaining all the possible diversity needed. The high opportunity costs for land with farmers' varieties is calling for action. Such a qualitative high level of agrobiodiversity in situ represents the optimum level of in situ conservation. But the large group of marginalized farmers are utilizing far too much land with farmers' varieties than is needed to maintain a national optimal level, because they have no alternatives.

Although the high opportunity costs are reason enough to reduce the area under traditional varieties as much as possible, the high fixed costs of approximately US $ 500 are questioning expensive in situ programs and projects. Hence, a system of in situ conservation has to be found which is flexible enough to react when needed and is less expensive.

As a first preliminary approach at estimating the costs of in situ conservation based on the introduced "controlled in situ conservation" system (see Chap. 3.4.2.2), the following details have to be considered:

For each traditional variety to be conserved in situ an average 1,000 m^2 is sufficient. For safety reasons, it was further recommended that 10 to 50 farmers should cultivate one distinct variety to reduce the risk of yield failure. Consequently, less than 5 ha are necessary for the in situ conservation of one distinct variety or landrace. These 5 ha is the areal Minimum Safety Standard (aMSS).

Furthermore, all agricultural crops consists of approximately 3 million distinct varieties (FAO, 1996a). Hence, an aMSS of roughly 15 million ha of arable land must be utilized for a safe, but minimized in situ conservation; 1% of the existing 1.4 billion hectares arable land (Engelman and LeRoy, 1995) is necessary to conserve the estimated 3 million distinct varieties[54].

As long as the economic and technological development has not yet transformed all marginalized areas into high potential areas, much more than 1% of the arable land is still utilized with traditional varieties and landraces. In India, more than 40% of arable land is cultivated with high-yielding varieties (CMIE, 1988). Hence, in most cases no intervention is required; no financial costs will be incurred. In situ conservation, however, is supposed to be sustainable even in the face of changes of comparative advantages of traditional varieties to modern varieties. Consequently, a rough estimate has to guide the decision whether in situ conservation will be justifiable in the light of scarce financial resources.

To quantify the overall costs for in situ conservation, a detailed analysis is needed on a country level. These analyses must take into account the degree of technological development in each region. The incentives are only justifiable if the technological development has opened the opportunity for farmers to change their production system and therefore the risk of a declining agrobiodiversity is evident. The in situ conservation has then only to be stimulated.

In the first place, financing the controlled in situ conservation may lay in the hand of each country, which has the sovereignty over its genetic resources, as was stressed by CBD (UNEP, 1994a). There may, however, be countries, which need more than the calculated 1% of their arable land for the in situ conservation of all

[54] For the sake of comparison: India has set aside 4% of the country's surface for almost 500 wildlife protected areas (Singh, 1996).

distinct varieties because of their richness in agrobiodiversity[55]. On the other hand, there may be countries, which need less than the average 1%. Because of the interdependency in PGRFA already utilized and needed in the future, the countries having to spend less than average may compensate countries which need more than 1% of their arable land.

India and Germany are examples of an agrobiodiversity-rich and -poor country; it can be stated that while in India land under agricultural use is still expanding to meet the increasing food needs, in Germany 12% of arable land is left as fallow due to the EC's set-aside programs (GCR, 1996). Consequently, the social costs on national level for a controlled in situ conservation, in terms of opportunity costs for foregone benefit, will be much higher in India than in Germany. From a simplistic point of view, the resources would seem allocated best if Germany could take over India's controlled in situ conservation. The different ecological conditions, however, restrict a transfer of in situ conservation activities from India to Germany. Consequently, Germany and other agrobiodiversity-poor countries could compensate India for the conservation which is in a global interest similar to the joint implementation programs, known from the UN Framework Convention on Climate Change.

5.1.3 Transaction Costs of PGRFA Conservation

The institutional and political negotiations and arrangements for developing and maintaining a system of PGRFA conservation and exchange can be seen as the system's transaction costs, as was defined by Coase (1960) and Williamson (1975). The following costs belong to specific transaction costs of the system of PGRFA conservation and exchange in its miscellaneous possibilities:

- the costs for the negotiations in the different fora (FAO, WTO, UPOV, and CBD),
- the costs for running any international and national approved PGRFA exchange system,
- the costs for the information and control management and
- the costs for any other activity related to the implementation and maintenance of the evolving system.

Mainly the costs emerging from international negotiations and arrangements are reflected in this study. Additionally, there are significant hidden transaction costs, mainly on a national level. Activities like the harmonization of national policies concerning the conservation and utilization of PGRFA between the different ministries, i.e., the agricultural ministry and the ministry for environment, trade or others, the development of one consistent system for a national program or institutional arrangements therefore could not be quantified in this study.

Another area of activities concerning conservation of PGRFA, even though with more indirect impact, are all activities for institution and capacity building and the

[55] See Chap. 3.1 for the concept of agrobiodiversity-rich and -poor countries.

creation of institutional legal frameworks. The international costs of such activities were calculated for "The Global Plan of Action for the Conservation and Utilization of Plant Genetic Resources for Food and Agriculture" to be a minimum of US $ 150 million (FAO, 1996g). If it is calculated that there are about 3 million distinct accessions (FAO, 1996a), which have to be conserved in ex situ as well as in in situ surroundings, the average costs for the institutional framework per distinct accession will be US $ 50.00 per year. These are the costs for supporting activities, e.g., institution and capacity building and the creation of institutional legal frameworks.

Transaction costs are generally costs for information extraction and production, monitoring, processing, coordination as well as follow-up costs of incomplete information, e.g., negotiation costs, enforcement costs etc. Following the above discussion, the following transaction costs can be defined for the establishment, running, and controlling of any exchange system for conservation and utilization of PGRFA:

- costs for the establishment of any exchange system
  - costs for negotiations (CGRFA, CBD, TRIPS, UPOV)
- costs for the running of any exchange system
  - costs for search, information procurement and processing (WIEWS)
  - costs for administration of mechanisms for benefit sharing
- costs for controlling exchange systems
  - costs for monitoring and enforcement of liabilities[56] .

5.2 Expenditures of PGRFA Conservation at Country Level

Besides costing PGRFA conservation according to the different categories, as was undertaken in Chapter 5.1, the costs can be estimated according to their origin. Considering the players in the conservation activities (see Chap. 4.2), costs can arise on farm, national and international level as well as on the level of conservation activities in the private sector. As part of the preparatory process for the International Technical Conference on Plant Genetic Resources in Leipzig, countries were surveyed as to their domestic expenditure on PGRFA related activities. The results are analyzed and discussed in this section.

[56] A cost analysis of molecular marker technologies to determine genetic origin, revealed running costs per data point ranging from US $ 0.14 to US $ 1.30. In Mexico, however, costs could reach US $ 1.72, due to higher prices for specific inputs. (Ragot and Hoisington, 1993).

National expenditures on PGRFA conservation are difficult to assess, largely because of uncertainties in defining the scope of PGRFA programs. It seems that most countries' national efforts to conserve PGRFA are in the hand of different departments in different ministries. In addition to the complex administrational structure, other parastatel and non-governmental organizations are involved in the conservation activities as well. Only in some countries are all efforts coordinated by an overall national program. Hence, the costs involved are not always visible. Furthermore, countries are involved in PGRFA conservation but do not have specially defined budget lines for these activities. For instance, if a genebank belongs to a national breeding institute and its costs are incorporated in the institute's overall budget[57], it is difficult to assess its specific costs.

In view of the inherent difficulties and limitations in compiling data on current expenditures on PGRFA, the purpose of these costing estimates were confined to obtain orders of magnitude of the current efforts made.

Information was requested from 154 countries at the end of 1995. Each country had established a focal point for the preparatory process of the International Technical Conference on Plant Genetic Resources. These focal points were contacted for this purpose. As of June, 1996, 43 countries had replied. Four of these countries provided data which could not be employed (Angola, Gabon, Guinea, and Zaire). Thus, 28% of all surveyed countries provided data which could be analyzed. Among those responding were countries thought to have substantial programs in PGRFA (inter alia, the USA, Germany, Russian Federation, China, India, and Ethiopia), as well as a number of countries with smaller programs.

The data concerning the national expenditures for conservation of PGRFA can be divided into three different groups:

- domestic expenditures, which have been spent for conservation activities in the country;
- expenditures, which have been spent for conservation activities in the country, but which derive from financial assistance received for PGRFA conservation (through bi- or multilateral contributions);
- expenditures, which have been contributed as financial aid for PGRFA conservation (through bi- or multilateral contributions).

The most important group for the actual purpose is the domestic expenditures. Countries receiving financial assistance may have domestic expenditures equaling the financial assistance, i.e., utilizing the financial assistance for conservation activities. But countries may spend more than the financial assistance only for conservation activities, i.e., these countries do utilize domestic financial resources for PGRFA conservation. Therefore, domestic expenditures include expenditures derived from received financial assistance. Table 5.9 indicates the nature of the information received from each country.[58]

---

[57] The genebank in Gatersleben/Germany, for instance, is part of the IPK.

[58] The difficulties concerning the interpretation of the data are:

1. The data received were not homogenous. Countries provided data based on different definitions of PGRFA conservation. Some countries included plant breeding activities,

The information received was tabulated into the main two cost categories (ex situ and in situ conservation activities), wherever possible. Difficulties arose in defining the scope of PGRFA expenditure. The informational content received by the countries was very different. Some countries prepared a more detailed analysis (e.g., Germany), some presented data on a very high aggregation level (e.g., USA). Therefore, the comparison of all received data is only possible on a high aggregation level, i.e., discussing expenditures for PGRFA conservation as a total. It is not possible to differentiate the expenditures made by the countries into the three different categories as described in Chap. 5.1.

Based on the data provided, the order of magnitude of domestic expenditures spent for the conservation of PGRFA by 37 of 43 named countries amount to approximately US $ 475 million for the year 1995 (see Table 5.9). This figure includes the financial assistance of US $ 17 million, which 15 countries received through bilateral and multilateral contributions. From the 39 countries mentioned above, 12 contributed bi- and multilateral financial assistance of approximately US $ 50 million.

The differences between the foreign assistance contributed by countries and that received by countries are determined by two facts:

- not all contributed assistance is distributed bilaterally, therefore the difference is partially determined by the multilateral financial contribution, e.g.,to international organizations which operate their own programs;
- furthermore, the difference is determined by the fact that the bilateral financial assistance of the respective 12 countries is not contributed exclusively to the 16 receiving countries mentioned here.

The interpretation of data, particularly highly aggregated data, is problematic. With more information from the countries, a more precise estimate of the overall expenditure could be made. Concluding the analysis of the international expenditures for the conservation of PGRFA, the survey result for 39 countries is drawn up in Fig. 5.3 ; 89% of the expenditures by the OECD countries surveyed (of US $ 456 million) are spent for their domestic conservation activities (US $ 406 million), mainly the ex situ conservation of their PGRFA accessions. 76% of the expenditures for the conservation activities in the developing countries surveyed (US $ 52 million) were funded nationally, nearly one quarter of the domestic expenditures were funded through bi- and multi-lateral financial contributions (US $ 16 million). Even though the 16 OECD countries are conserving 53% and the 23

---

whereas others only included the conservation of PGRFA in a very narrow sense. Most countries did not clearly define what was covered by expenditure or foreign assistance data. Similarly, some data on assistance contributions included only activities closely related to conservation and utilisation of PGRFA. Some countries provided data only on the general national contribution to international organisations.

2. The data received were often not desegregated. Sums were given without indicating whether it applied to conservation, utilisation, or both.
3. Some countries, although they probably contribute foreign assistance for PGRFA, did not send any data on the amount of contributions.
4. Other countries, which probably receive foreign assistance did not provide data on this.

developing countries 47% of their combined conserved accessions, the OECD countries spent 85% of the their combined total of US $ 475 million. Not surprisingly, the contribution for the international activities originate predominantly from the 16 OECD countries.

When the countries are grouped into agrobiodiversity-rich and -poor countries and furthermore into countries having high and low absolute domestic expenditures for PGRFA conservation, most of the OECD countries can be found among the agrobiodiversity-poor countries with the tendency to higher absolute expenditures (see Fig. 5.4)[59]. The majority of these genetic resource-poor countries are very much interested in building up and maintaining a high level of PGRFA collection from all different countries and gene centers to supply their breeding industry with sufficient resources and to ensure a long-term sustainable food production.

Fig. 5.3. Expenditures for the conservation of PGRFA by 39 surveyed countries (in US $ million)

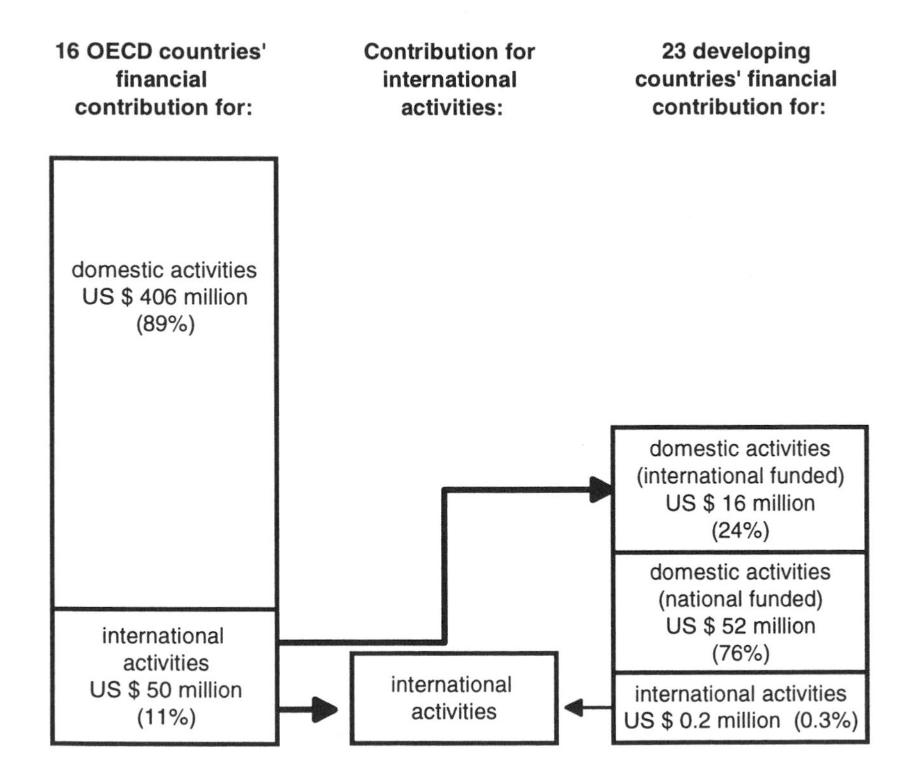

[59] See Chap. 3.1 for the concept of agrobiodiversity-rich and -poor countries.

Table 5.9. National expenditures for PGRFA conservation

Country	Domestic expenditures 1995 (incl. foreign received assistance) in US $	Foreign assistance contributed 1995 in US $	Foreign assistance received 1995 in US $
Germany	113,215,000	18,527,000	
France	98,660,000	500,000	
United Kingdom	70,154,000	17,531,000	
Spain	33,413,000	885,000	
Italy	27,208,000	n.i.	
United States of America	20,433,000	n.i.	
South Africa	19,000,000	n.i.	n.i.
Norway	16,208,000	2,612,000	
Egypt	11,528,000	244,000	3,551,000
Greece	10,958,000	n.i.	
Brazil	8,000,000		1,112,000
India	6,776,000		4,565,000
Japan	6,480,000	n.i.	
Peru	4,137,000		1,883,000
Switzerland	3,825,000	3,400,000	
Slovak Republic	3,608,000		
Czech Republic	3,255,000		
China	2,526,000		1,350,000
Madagascar	2,385,000		1,577,000
Seychelles	2,322,000		302,000
Haiti	1,896,000		796,000
Canada	1,584,000	3,580,000	
Russia	1,526,000		260,000
Ethiopia	1,346,000		969,000
Portugal	1,030,000	10,000	
Suriname	1,028,000		778,000
Poland	656,000		10,000
Lesotho	615,000		172,000
Romania	408,000		
Tanzania	187,000		n.i.
Cyprus	186,000		7,000
Togo	151,000		n.i.
Belarus	135,000		n.i.
Pakistan	120,000		n.i.
Tonga	56,000		26,000
Saint Kitts and Nevis	20,000		n.i.
Austria	10,000	1,500,000	
Finland	n.i.	1,180,000	
Ireland	n.i.	142,000	
TOTAL:	475,045,000	50,111,000	17,358,000

Source: Data according to the questionaires and interviews with respective focal points.

Additionally, a country like Egypt may be seen as an example for those genetic resource-poor countries, which still has a large agricultural sector. Agriculture in these countries has to be undertaken under harsh conditions however. Consequently, the need for the sustainable supply of crucial inputs must be taken care of.

Fig. 5.4. Absolute domestic expenditures for PGRFA conservation for selected countries

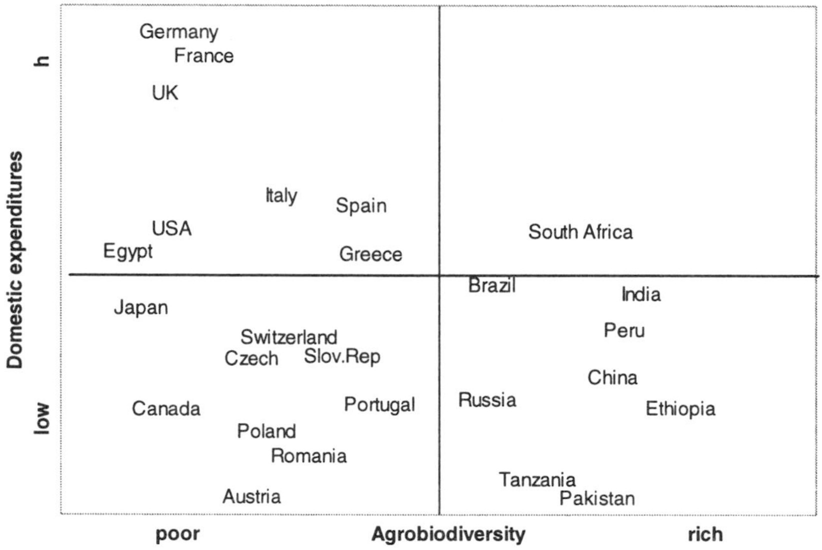

Domestic expenditures
h
low
Germany
France
UK
Italy
Spain
USA
Egypt
Greece
South Africa
Brazil
India
Japan
Peru
Switzerland
Czech
Slov.Rep
China
Canada
Portugal
Russia
Ethiopia
Poland
Romania
Tanzania
Austria
Pakistan
poor
Agrobiodiversity
rich

Note: low domestic expenditures: less than US $ 10 million for PGRFA conservation;
high domestic expenditures: more than US $ 10 million for PGRFA conservation;
agrobiodiversity-poor country: country is not part of a gene center or has less than 10,000 accessions stored ex situ;
agrobiodiversity-rich country: country is part of a gene center and has more than 10,000 accessions stored ex situ.
For the presentation, the domestic expenditures for PGRFA conservation underlie a logarithmic scale.

Source: calculation based on data from the respective countries (see Appendix 13)

Even if the domestic expenditures are expressed as percentage of the GDP/capita, a country like Egypt still has high ranking as regards its expenditures (see Fig. 5.5). In this case, however, it can be seen clearly that not only resource-poor countries are interested in the conservation of PGRFA, but some agrobiodiversity-rich countries as well - India, Ethiopia, South Africa, China, and Tanzania. These countries are spending as much for PGRFA conservation in

[60] See Chap. 3.1 for the concept of agrobiodiversity-rich and -poor countries.

relation to their average income as resource-poor countries like Germany, France, and the UK. Especially in India, Ethiopia, and China the expected value for PGRFA conservation is assessed as being very high. They are also taking a leading role in the negotiations for the internalization of or the compensation for the maintenance of PGRFA in their countries.

Fig. 5.5. Relative domestic expenditures for PGRFA conservation for selected countries

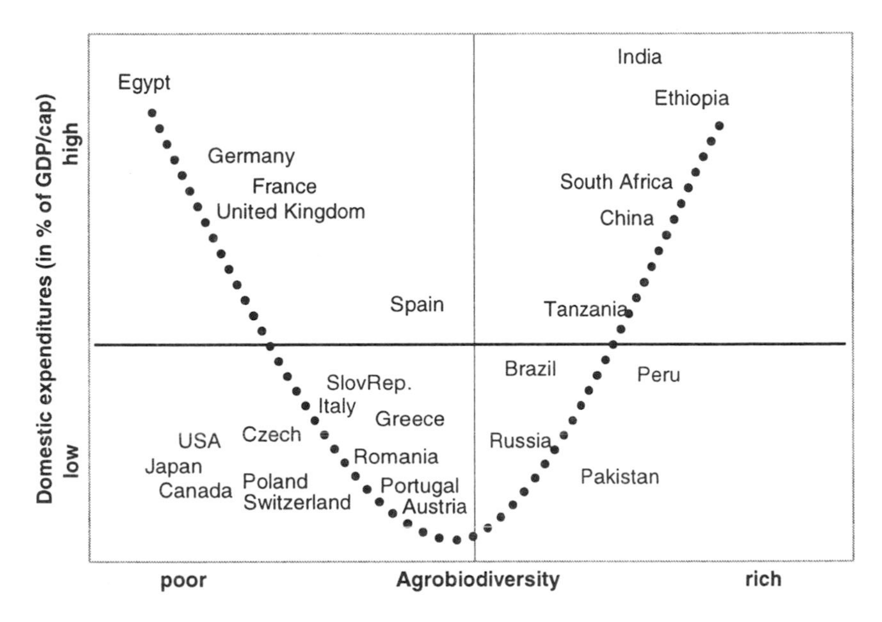

Note: low domestic expenditures in % of GDP/cap: less than 200% of GDP/cap for PGRFA conservation;
high domestic expenditures in % of GDP/cap: more than 200% of GDP/cap for PGRFA conservation;
agrobiodiversity-poor country: country is not part of a gene center or has less than 10,000 accessions stored ex situ;
agrobiodiversity-rich country: country is part of a gene center and has more than 10,000 accessions stored ex situ.

Source: Calculation based on data from the respective countries (see Appendix 13)

On the contrary, however, there are countries, resource-poor as well as resource-rich, which do not spend very much for the conservation efforts in relation to their average income. Genetic resource-poor countries with low financial commitment are mainly countries with few or no activities in the breeding and seed industry. On the other hand, a country like Pakistan does not invest much for a conservation program in spite of being genetically resource-rich.

Except for some countries, like the USA, Canada, and Japan, with good national conservation systems, but low public commitment in the further steps of processing the genetically coded information leading to plant breeding, the countries are grouped like a broad-shaped "U" - its two ends characterizing countries strongly committed to PGRFA conservation, based on opposite prerequisites. On the one side (left top) are the demand-driven spenders. These are agrobiodiversity-poor countries which spend much for PGRFA conservation. The governments of these countries feel the need for their breeding industry to safeguard their demand for genetic resources as input for breeding. On the other side (right top) are the supply-driven spenders, which are agrobiodiversity-rich countries. These countries invest much for the conservation of PGRFA. They do so for their own country's breeding efforts but also to be able to appear on a market for genetically coded information yet to be developed.

So far, the expenditures for PGRFA conservation were interpreted as similar qualitative investments into an effective conservation system. The cost-effectiveness of conservation has not been discussed at all yet. For the time being, a simple, but ultimately, the most important input-output ratio is deduced from the information gathered. The domestic expenditures for PGRFA conservation are compared with the numbers of accessions stored in the respective country in Table 5.10. The ratio gives a first impression on the efficiency of conservation operation in the selected countries. The countries are divided into three groups, based on the unit costs of conserved accessions in the country.

It can be seen that countries with the same amount of conserved accessions may have very different unit costs. With 17,000 accessions, Switzerland pays US $ 225 after all to conserve one accession, whereas Pakistan, which conserves approximately 19,000, has unit costs of only US $ 6.

Besides the differences in conservation quality, the differences in conservation activities included in the overall expenditures reported by the countries determines mainly the varying ratio. For instance, the different approach in plant breeding has to be understood in order to explain the big difference between USA's unit cost of US $ 37 and the high ratio for some countries of the European Union. In the USA, PGRFA conservation is strictly restricted to the activities of collection, storage, regeneration, multiplication, and documentation of the germplasm (Shands, 1994). In Germany, the Braunschweig Genetic Resources Center (BGRC), for instance, is integrated in the Federal Center for Breeding Research on Cultivated Plants (BAZ) and evaluation and genetic enhancement are combined research activities.

An order-of-magnitude estimate of expenditures spent by all countries in 1995 was carried out based on data received from the 39 countries. 165 countries were integrated in the estimate. Such countries were ignored that did not participate in the preparatory process for the International Technical Conference on Plant Genetic Resources, or did not have any kind of national program for the conservation of PGRFA, and where no information is available on any PGRFA conservation activities in the respective country. In such a country, the possibility that this country has any expenditures for PGRFA conservation activities is reduced to zero[61].

---

[61] Therefore countries like Kuwait, Djibouti, Singapore, and Liechtenstein were excluded. Furthemore, the Democratic People's Republic of Korea (North-Korea) was excluded because of unsafe information. (For a full list of countries participating, see Appendix 12.)

Besides the countries chosen, some criteria had to be identified which are assumed to have the most relevant and significant correlation between national conservation activities and the conservation expenditures in different countries.

Table 5.10. Ratio of domestic conservation expenditures to ex situ conserved accessions for selected countries

Country	Domestic expenditures in 1995 (in US $,000)	Accessions stored ex situ	Ratio conservation expenditures – conserved accessions (in US $)
Russian Federation	1,526	333,000	5
Pakistan	120	19,208	6
Canada	1,584	212,061	7
China	2,526	350,000	7
India	6,776	342,108	20
Ethiopia	1,346	54,000	25
Japan	6,480	202,581	32
United States	20,433	550,000	37
Brazil	8,000	194,000	41
Tanzania	187	2,510	75
Switzerland	3,825	17,000	225
Italy	27,208	80,000	340
Germany	113,215	200,000	566
United Kingdom	70,154	114,495	613

Two different approaches, based on different criteria utilized, were used to obtain results of the estimate:

1. Gross Domestic Product: For the first approach, GDP was taken as criterion. The overall expenditures were estimated by correlating the overall GDP of all 165 countries to the sum of the specific countries' GDP and multiplying it with the known expenditures taken from the 39 countries (for financial assistance received, contributed, and domestic expenditures respectively).[63]
2. Accessions: Based on the assumption that most of national expenditures of the conservation of PGRFA is made in the area of ex situ conservation, the ex situ

[63] For the detailed calculations see Appendix 14.

conserved accessions were used as criterion in the second approach. Calculating the ratio of all ex situ conserved accessions to the accessions conserved by the 39 countries and multiplying it with the known expenditures of these countries, the overall expenditures for the three categories was estimated.[63]

Because of the significant different expenditure structures in different countries, as some examples demonstrate in Table 5.10, the countries were clustered into different groups. Without clustering the countries, the estimation would become distorted in one way or the other. For instance, financial assistance contributed for international and national activities in PGRFA conservation is mainly carried out by OECD countries. Without any kind of clusters, the estimate would imply financial contributions by all countries, which overrates these contributions. Consequently, all 165 countries were clustered into different groups according to the following criteria:

Clustering countries by *GDP per capita* takes into account that countries with similar economic power may spend equal amount for PGRFA conservation. The countries were divided into 5 groups. In order to represent the economic resemblance more roughly all countries were divided into two groups: the OECD and the non-OECD countries. Another way of clustering was to group the countries into 6 groups according to their *region*, which may present the similarity in the economic conditions between the countries in a different but also significant way. A third way of clustering the countries was by dividing the countries into 3 groups according to *accessions* and *GDP per capita* according to the numbers of accessions conserved ex situ (high, medium, and low). Each group was subdivided into 3 groups according to their GDP per capita.

It is necessary to identify plant genetic resources rich and poor countries when estimating the future costs for PGRFA conservation, according to criteria as the amount of accessions stored in ex situ collections (as key criterion for the national situation of ex situ conservation activities) as well as the fact whether the specific country is part of one of the gene centers (as key criterion for the national situation of in situ conservation necessities). But estimating the expenditures for 1995 has to be mainly derived from the ex situ conservation activities, because only a few countries have started with in situ conservation activities for PGRFA and may already have significant expenditures in 1995. Consequently, for the estimate carried out here, only the numbers of accessions stored in genebanks in a country are implied to cluster the countries in groups with similar conservation activities and expenditures.

By using the different approaches and criteria noted above, and by analyzing the results, the estimate may be considered to be the most reasonable, given the data constraints already mentioned.

All the estimate results are shown in Table 5.11. Their most important aspects will be discussed in the following[64]. Based on the data provided and the different calculations carried out, it is estimated for 1995 that the total of the national

[63] For more details of the calculations see Appendix 14.

[64] The detailed calculations can be looked up in Appendix 14.

domestic expenditures for conservation of PGRFA were approximately US $ 733 million. Depending on the different approaches used, however, the order of magnitude is between US $ 599 million and US $ 740 million. These results represent a variation of approximately 10% from the average value of US $ 674, or a deviation of 1% over and 18% below the most likely value of US $733 million. Based on the information received, it is estimated that the group of 24 countries, mainly from North America and West Europe with the highest GDP per capita account for approximately US $ 512 million of the total, whereas all other countries combined account for US $ 221 million of the overall national expenditures for PGRFA conservation.

The *domestic expenditures* spent for PGRFA conservation can be calculated by clustering the countries into 6 regional groups and correlating the expenditures to the amount of accessions, (approach 1 in Table 5.11). The overall economic situation in each regional group is similar enough to calculate average domestic expenditures per conserved accession. Therefore, this approach is preferred for the estimation of the overall domestic expenditures, amount to US $ 733 million in 1995. The same neglectable distortion is more or less assumed to be existent by dividing the countries according to the number of accessions and GDP/capita (approach 2d in Table 5.11). The other three approaches (2a, b, c in Table 5.11) are assumed to have some distortion because of under-estimating the low income countries' potential of domestic expenditures.

The order of magnitude in the *financial assistance received* by countries is high, ranging from US $ 58 million to 164 million, because only 10% of all relevant countries which potentially received financial assistance for the conservation of PGRFA presented data. Correlating financial assistance to the number of accessions stored ex situ and simultaneously clustering the countries into 6 regional groups represents the best possible solution for the estimate (see approach 1 in Table 5.11). The overall economic situation in each regional group is similar enough to calculate the average financial assistance received per conserved accession. For each group there are enough data from countries representing the group. Consequently, the figure of US $ 82 million is the closest rough estimate of financial assistance received by countries for thr conservation of PGRFA in 1995. The clustering of countries according to their OECD membership (approach 2a in Table 5.11) integrates most of the countriese contributing financial assistance on the one hand and those countries receiving financial assistance on the other.

Because of lacking information the calculated US $ 58 million are probably underestimated. The other two approaches (2b and 2c in Table 5.11) overestiomate the financial assistance by integrating countries which contribute financial as receiving countries. Because little information for each group in the category of financial assistance was received, the approach of clustering the countries into 9 groups (approach 2d in Table 5.11) becomes distorted[66].Consequently, this figure is not presented in the overview in Table 5.11.

Some studies commissioned by UNEP estimated the financial resources provided to developing countries for the conservation of biodiversity in general. Their result

[66] For detailed information see Appendix 14.

was that less than US $ 240 million were contributed annually (Zedan, 1996). This result proofs that the estimations presented in this study are in an acceptable range of magnitude, based on the assumption that far more financial resources are contributed for the conservation of biodiversity in general than it is for agrobiodiversity. The approach 2a in Table 5.11, clustering the countries according to the OECD membership and correlating the financial assistance contributions to GDP, presents the best approach for the estimate of the *financial assistance contributed* by countries through bi- and multi-lateral channels.

Table 5.11. Estimation of expenditures for the conservation of PGRFA according to different criteria

Approach	Estimation Criteria	Cluster Criteria	Estimated foreign assistance received for PGRFA conservation 1995	Estimated foreign assistance contributed for PGRFA conservation 1995	Estimated domestic expenditures for PGRFA conservation 1995
				[US $ '000 000]	
1	Accessions	Grouped into 6 regional groups	82	---	733
2a	GDP	Countries grouped into OECD and non-OECD countries	58	189	599
2b	GDP	Countries grouped into 5 groups according to GDP/capita	107	176	642
2c	GDP	Countries grouped into 6 regional groups	164	166	655
2d	GDP	Grouped into 9 groups according to Accessions and GDP/capita	---	164	740

See Appendix 14 for detailed calculations.

Most of the potentially contributing countries are grouped in one group and all the potential recipients are in the other group. Therefore, US $ 189 million is the closest estimate possible. The three other approaches (2b, c, and d in Table 5.11) differ from approach 2a in Table 5.11 by less than 13% and hence confirm this result. The estimation according to approach 1 in Table 5.11 is not relevant, because the contributions will not be significantly correlated with the nationally stored accessions.

Limitations of data and methodology inevitably produce bias when formulating estimates such as those presented here. It should be noted that:

- The estimate cannot be more precise than the underlying data. The above mentioned limitations of interpretation should also be taken into consideration.
- The estimate of domestic expenditures based on estimation of the data provided is reasonable, because information was available from 28% of all countries, including countries with major as well as more modest programs. As noted earlier, the scope of domestic expenditures is, however, not clearly defined.

5.3 Analysis of the International Expenditures in 1995

The survey of countries and the estimate indicate that about US $ 160 - 190 million dollars are contributed each year for activities relating to PGRFA conservation on an international level as well as financial assistance to specific countries (see Table 5.11)[67]. This figure is likely to be underestimated because the surveyed countries contributing financial assistance did not always indicate all their contributions, especially not the bilateral contributions. Furthermore, it is difficult for the countries to judge how high the percentage of their contribution is to international agencies, using it for activities related to PGRFA conservation. For instance, all CGIAR members have given their specific contributions to the system, but only a part of the CGIAR's activities may be ascribed to PGRFA conservation.

A significant number of international funding and executing agencies are involved in activities relating to the conservation of PGRFA. Consequently, in addition to national activities there are significant efforts undertaken on international level to conserve the diversity of PGRFA. In the process of this study, a survey was made of the expenditures on an international level. Fig. 5.6 depicts the flow of financial resources for conservation of PGRFA, which may be divided into two different groups. On the one hand, there are activities by international institutions and organizations, e.g., FAO or the CGIAR centers. Single countries or all countries benefit from the output of their work. This may lie in the access to unique accessions, conserved in one of the CGIAR centers' genebanks or it may lie in specific programs and projects implemented by FAO or other implementation agencies in single countries or specific regions. All these activities are financed by the contributions of member countries to the different organizations and institutions. Besides the organizations mainly implementing programs and projects, which also include NGOs, there are international organizations in charge of funds for contributing financial assistance, e.g., GEF and the regional development banks. Their contributions, either as grants or credits, may be implemented by national organizations or by international implementing organizations. The World Bank and

[67] The expenditures for international activities as well as financial assistance to specific countries through international channels are called international expenditures in the following.

other development banks and funds are major players in agricultural development projects and NARS capacity building. However it is difficult to assess the proportion related to PGRFA. Hence, international flows include about US $ 7 million annually channeled through the Global Environment Facility (GEF) for PGRFA related activities[67] (Virchow, 1996a). Thus only a small portion of the GEF currently allocated to PGRFA can be estimated.

On the other hand, the financial assistance contributed by some countries for specific conservation activities in other countries also play an important role. To quantify these contributions is difficult. For instance, the new genebank in New Delhi, India, which was inaugurated in November 1996, was financed mainly by the USA. The USA, however, did not mention the US $ 28 million nine year program with India, even though there were significant expenditures for the project in 1995 (Chandel, 1996).

Expenditures of funding and executing agencies for conservation of PGRFA were calculated by compiling the information gathered from communications and documents beginning in September 1995 up to early 1996. The information received represents most of international agencies which are involved in activities relating to conservation of PGRFA (see Table 5.12). Furthermore, the risk of double-counting financial resources was eliminated in this approach by tracking the different activities and projects funded by cross checking the project codes. In addition to this, the differentiation between the national expenditures, including the listing of financial assistance contributed to other countries or national and international agencies, and the expenditure estimation on international level, eliminates the risk of double-counting.

Bearing the limitations of this approach in mind, the rough overall estimate of the expenditures in PGRFA conservation on international level is summarized in Table 5.14 and can be described as follows.

About US $ 83 million were channeled through international organizations to activities relating to the conservation of PGRFA in 1995. Most of the money was spent by the above mentioned international funding and executing agencies for the conservation of PGRFA, mainly as an integrated element of larger development projects or programs.

However, information was not available for some of the agencies listed above. Nineteen major NGOs active in the field of PGRFA with specifically programs covering all regions were surveyed, of which one responded. Similarly, information from private foundations is scant.

---

[67] This includes two projects specifically devoted to PGRFA, plus an estimated share (5%) of a number of projects for biodiversity conservation which are likely to contribute to PGRFA conservation.

Fig. 5.6. Flow of funds for the conservation of PGRFA

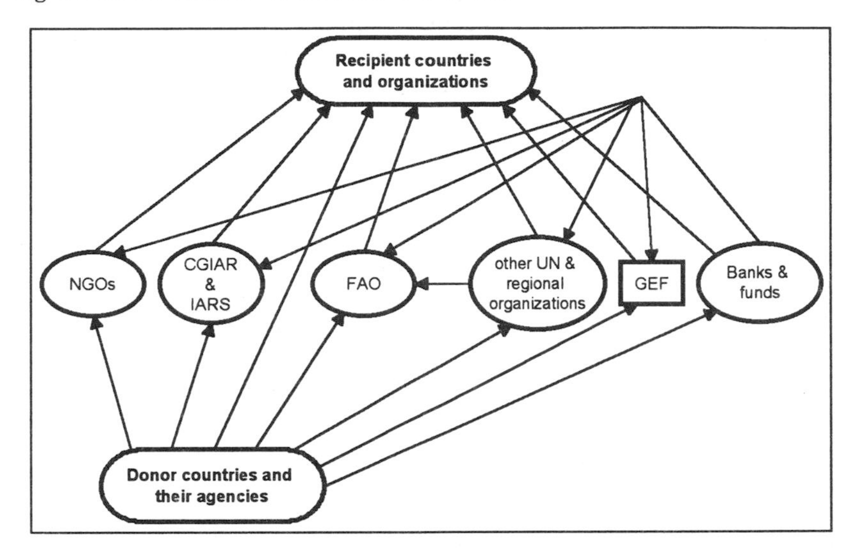

Three basic assumptions were made in estimating the amount of financial resources spent on the international level for conservation of PGRFA in 1995:

1. An agency's projects or programs often dealt with PGRFA conservation as part of a broader initiative including actions not strictly related to PGRFA. In these cases the portion of the program dealing with conservation of PGRFA had to be estimated as a percentage of total. The impact range of each project was estimated on the basis of the project's information into 4 categories: small (which estimated a 5% project's proportion on PGRFA conservation), medium (25%), large (50%), and absolute (100%).
2. Furthermore, it was assumed that the proportion of PGRFA-related costs of the total project or program costs were equal to the proportion of the share on conservation of PGRFA as described by the agency (e.g., if a project was estimated to have a 25% share on conservation of PGRFA, 25% of the overall costs of the project were calculated). It is obvious, however, that the real costs for the PGRFA conservation activities may be much smaller (or higher) than the project's proportion on PGRFA conservation.
3. Lacking more specific data, expenditures for projects and programs lasting for more than the year 1995 were averaged to obtain an estimate of 1995 costs [e.g., if a project were in operation from 1993 to 1996 (4 years) with total budget of US $4 million, the annual average expenditure would be calculated as US $ 1 million and applied to 1995].

Furthermore, the risk of double-counting financial resources was eliminated in this approach by tracking the different activities and projects funded by cross checking the project codes. In addition to this, the differentiation between the national expenditures, including the listing of financial assistance contributed to other countries or national and international agencies, and the expenditure estimation on international level, eliminates the risk of double-counting.

Bearing the limitations of this approach in mind, the rough overall estimate of the expenditures in PGRFA conservation on international level is summarized in Table 5.13 and can be described as follows.

About US $ 83 million were channeled through international organizations to activities relating to the conservation of PGRFA in 1995. Most of the money was spent by the above mentioned international funding and executing agencies for the conservation of PGRFA, mainly as an integrated element of larger development projects or programs.

Only a minor part of the whole international expenditures, namely US $ 7 million, has been spent by the technical assistance of mainly FAO as well as UNDP and UNEP financed projects, which represents 8% of all expenditures.

These expenditure estimates include the CGIAR's expenditures. The CGIAR's expenditures on PGRFA related activities represent with approximately 60% the biggest single expenditure share in all of the international expenditures (including core funds and complementary activities). The expenditures for the CGIAR genebanks is estimated at US $ 50 million[68]. The rest of the expenditures for PGRFA conservation, US $ 26 million comes from funds, such as the World Bank and GEF. Because of the lack of information on the projects and programs of the World Bank, some other development banks, and IFAD, it was not possible to include an exact figure derived from the survey in the above totals. Therefore, the expenditures for two of the main funding organizations were estimated. It was assumed that 0.5% of the US $ 3 billion of WB's agricultural expenditures and 2% of IFAD's 200 million budget were spent for activities related to PGRFA conservation. GEF's spending is determined by all their projects and programs which had an impact on the conservation and utilization of PGRFA. These were tabulated and calculated according the above mentioned method. The same procedure was utilized to calculate the expenditures of FAO, UNDP, UNEP, and CfC.

The data must be carefully interpreted, because most agencies do not have a separate budget for funding PGRFA activities. Because these activities make up only a small part of overall development projects and programs, it is difficult for each agency to calculate their expenditures for this task. Consequently, the estimates calculated on the basis of the received data can only be accepted as a first rough estimate and are subject to distortions because of the lack of accuracy of the primary data.

Concluding the analysis of the international expenditures for the conservation of PGRFA, the aggregated global expenditures can be drawn up by incorporating the estimated results of the international expenditures and the national expenditures contributed by countries in Chapter 5.2 (see Fig. 5.7).

According to the estimation, countries contributed US $ 733 million for national activities related to the conservation of PGRFA in 1995 (see Table 5.11). Additionally, US $ 189 million were spent for multi- and bilateral international

[68] A detailed calculations of the CGIAR's expenditures is given in Appendix 11.

activities (see Table 5.11). From this sum, the OECD's member countries have contributed approximately US $ 162 million. According to the estimation of international expenditures for conservation and utilization of PGRFA (including all technical assistance, funds and other activities of the international organizations included in the analysis) approximately US $ 83 million of the US $ 189 million were contributed through multilateral channels (see Table 5.13).

Table 5.12. Information from funding and executing agencies

Organizational category	Funding and executing agencies	Funding category
FAO	Regular budget plus TCP (Technical Cooperation) GCP (Government Cooperative Program): FFHC (Freedom from Hunger Campaign) UTF (Unilateral Trust Funds):	Technical assistance
Other relevant UN organizations	UNDP (United Nations Development Program) UNEP (United Nations Environment Program)	Technical assistance
Regional and other Agencies	CfC (Common Fund for Commodities) International Networks additional EU Programs Foundations Non-Governmental Organizations	Technical assistance
Banks and Funds	GEF (Global Environment Facility) other specialized UN and other trust funds (AGFUND (Arab Gulf Program), IFAD (International Fund for Agricultural Development) World Bank Group Regional Development Banks (African Development Bank, Asian Development Bank, European Bank of Reconstruction, Inter-American Development Bank, Islamic Development Bank)	Funds
IARS (International Agricultural Research Centers, including CGIAR and Non-Associated Centers)	All related centers	Research centers

[69] A detailed calculations of the CGIAR's expenditures is given in Appendix 11.

Table 5.13. International expenditures for the conservation of PGRFA

Origin of expenditures	Expenditures according to received information in US $ '000 000	Expenditures according to received information and some estimations in US $ '000 000	Share of expenditures to total
Technical Assistance	7	7	8%
Funds[a]	7	26	32%
Sub Total	14	33	
IARS	50	50	60%
Total:	64	83	

[a]: IFAD's and the World Bank's contributions were estimated
Source: Virchow, 1996a

Following the information received, approximately 60% (US $ 50 million) were contributed to international institutions for the conservation of PGRFA on the international level and approximately 40% (US $ 33 million) were contributed to international funding and implementing agencies for assistance to countries. Stressing once again the inaccuracy of the data, it is, however, still possible to obtain a tendency and a rough estimation of the current expenditures for PGRFA conservation, which amounts to over US $ 800 million.

By the same process of estimation, the foreign assistance received for conservation of PGRFA has been estimated to be approximately US $ 100 million. Because of the lack of information, there is no reliable estimation for the contribution through bilateral channels. From the known information, bilateral contributions are calculated to be at least US $ 106 million. The estimation of the foreign assistance received by countries would have to be increased by the unknown amount of contributions through bilateral channels to arrive at the total estimate.

Very little information is available on the conservation expenditures of the private sector and of the NGOs. Because of the lack of any information, no expenditures were estimated for the NGO's activities. Conservation costs for breeding companies concentrate on specialized ex situ activities. In addition to their elite germplasm, breeding companies store wild species, farmers' varieties and obsolete varieties as a stock for future breeding activities. The costs involved are maintenance and processing costs as well as partly acquisition costs, as far as the companies have collected the germplasm on their own.

According to the ASSINSEL survey, the members estimated their total costs of maintaining genetic resources to be about US $ 50 million per year (CGRFA, 1997). The figure is based on a study, which was carried out by ASSINSEL in 1996. The study was, however, not available upon request. Consequently, the figure could not be analyzed. According to WIEWS, the private sector stores approximately 1.3% of all accessions, which represent approximately 80,000 accessions. It follows that the ratio of conservation expenditures to ex situ conserved accessions is US $ 630, which makes it the highest ratio and is not very attractive financially. The high ratio can most likely be explained by assuming that the study included expenditures for breeding activities which are not plain conservation activities.

Additionally, 80% of their member companies are participating in national programs and 31% in international programs for conservation and utilization of PGRFA, whereby the financial participation was higher than US $ 1.5 million in 1996 (CGRFA, 1997). The total flow of official development assistance has been falling in recent years, as has the share of agriculture in total development finance. After the food crisis in the early 1970's annual commitments rose to around US $ 10 billion (in current value). This increased to about US $ 14 billion by the end of the 1980s. There has, however, been a steady decline since then, falling to around US $ 10 billion at present (UN, 1995). So, if financial assistance is assumed to be approximately US $ 190 million, it makes up 2% of the overall ODA for agriculture.

Fig. 5.7. Current national and international expenditures flows for the conservation of PGRFA in 1995 (in US $ million)

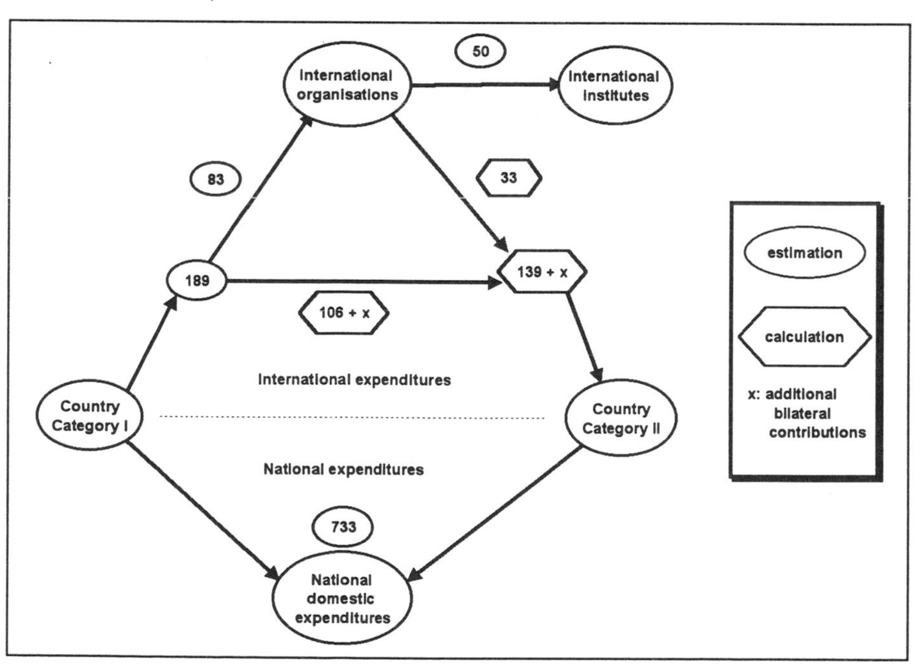

Note: Country I: a country which does not receive (international) financial assistance for its PGRFA conservation
Country II: a country which receives (international) financial assistance for its PGRFA conservation

A major part of the international expenditures represents the conservation efforts of the CGIAR System. The genebanks of the 12 CGIAR centers may be understood as the conservation sites which store the most utilized accessions at present, because of the direct interaction between conservators and breeders in the centers. Consequently, the costs for the CGIAR's conservation activities seem to be the most efficiently utilized financial resources for PGRFA conservation on international level. Furthermore, it is interesting to note that the bilateral financial assistance seems to be significantly higher than the multilateral assistance. In the future, this bilateral financial assistance will be the key focus point in negotiations and agreements. If the international negotiations on the access to and the benefit sharing of PGRFA makes progress, the outline agreement may be to compensate the access to PGRFA by increased financial assistance. Consequently, the flow of financial contributions will increase, partly via the international organizations, but mainly through the bilateral channels. The main question will be, whether additional contributions will proportionally improve the conservation efforts on national level.

So far in Chapter 5, the expenditures of conservation activities were examined for 1995. The question has not yet been discussed, whether the efforts undertaken are sufficient to achieve the discussed objectives (see Chap. 2.4).

According to the analysis of the Country Reports and the report on the state of the world's PGRFA (FAO, 1996a), maintaining the status quo would lead to a further degradation of genetic resources, in situ as well as ex situ. The analysis of the effectiveness of PGRFA conservation in some selected countries confirms the report (see Chap. 5.4). Consequently, more efforts need to be undertaken on a national and an international level to maintain the diversity of agricultural crops. The global plan of action (GPA), as outlined in FAO (1996b) and adopted at the International Technical Conference on Plant Genetic Resources in Leipzig, 1996, takes up all additional efforts needed for the conservation and sustainable utilization of PGRFA.

Drawing up and implementing a global plan of action with an incorporated global program may avoid duplicated efforts, add value through the creation of synergies among the partners, set global accepted priorities, and improve the access to relevant information (Frison et al., 1997). The benefit of the global plan of action is the increase of the efficiency of all efforts made to conserve genetic resources for food and agriculture on a global level. In the light of limited financial resources for all conservation activities, the increased efficiency through a global plan of action will ensure the best possible solution to conserve PGRFA.

In contrast to the GPA, this study only identifies the major needs having an impact on an increased degree of achievement for the above mentioned conservation objectives (see Table 5.14). These additional efforts can be divided into four different groups, which have impacts on the different objectives:

1. The activities of the first group aim at filling the gaps in existing ex situ collections as well as improving their storage security. Consequently, activities like surveying, inventorying, and collecting PGRFA will ensure that all varieties are conserved in ex situ collections as well as improve the accessibility of the varieties. Furthermore, activities like sustaining existing ex situ collections,

including the duplication of collections, and the regeneration of threatened ex situ accessions will secure the ability to store the accessions long term.

2. The purpose of the activities in the second group is mainly the improvement of the availability of accessions by expanding the characterization and evaluation activities as well as developing and improving the information and early warning systems.
3. Activities like on-farm management and improvement of PGRFA, in situ conservation of wild crop relatives and wild plants for food production, and the reintroduction of PGRFA into agricultural systems are aiming at controlling the genetic diversity in farmer's fields. These activities represent the third group and contribute to improved adaptation possibilities of an increased amount of varieties.
4. The final group summarizes supporting activities, like supporting or establishing institutions, mainly national programs and networks. Furthermore, this group includes activities enabling the capacity building in the countries. All the activities of this group have an impact on all three different objectives.

These additional efforts described above are understood as activities which further the conservation objectives defined. Therefore, the following costs analyzed are understood as incremental costs, i.e., costs which will not replace other costs already calculated (King, 1993). The baseline against which the incremental costs have to be defined is composed of the activities and their quality already under way, based on the 1995/96 information.

The activities listed include activities that may be funded by national governments, and other domestic sources of funds, as well as internationally through multilateral organizations, and from bilateral and regional sources. Only those costs are included in the costing, which are ascribed to by the international community. This includes a significant share of the costs of implementing activities in developing countries. It also includes activities undertaken largely for the global benefit, regardless of their location.

The costs have been estimated by aggregating known or forecast costs for the specific activities. Estimates can, however, not be made using detailed budget line items of formal submissions or project documents, since specific and detailed projects and programs have not yet been formulated. The results, presented as average annual costs during a 10-year period, at 1996 costs, should therefore be regarded as order of magnitude estimates only. Referring to the descriptions of the activities in the Global Plan of Action for the Conservation and Sustainable Utilization of Plant Genetic Resources for Food and Agriculture, these calculations are based on the cost estimates for the global plan of action and are recalculated according to the specific cases discussed here (see FAO, 1996b and FAO, 1996g).

As can be seen in Table 5.15, the estimates of the total cost for the essential activities to increase the effectiveness of the global conservation efforts amount to US $ 337 million. The results of the costing estimations indicate three potential levels, because of significant variables in the potential rate of implementation. Depending on the speed and intensity with which activities should be implemented, the cost of components of the calculation can thus vary by a large amount. For these reasons, the costs have been calculated, in each case, for three options.

Option A represents a basic or rudimentary approach to the implementation of the activities. In situations where costs could not be determined precisely, it assumes the least costly of reasonable possibilities. The number of countries, institutes or communities covered is lower than in other options. US $ 109 million were estimated for the basic approach.

The moderate approach, for which US $ 185 was estimated, is based on intermediate assumptions regarding needs. Its general coverage is greater than in the basic approach, but is consistent with known and documented needs and realistic absorption and implementation capacity of countries. US $ 337 million were estimated for a high-urgency approach. This approach reflects a very high necessity of activities at present and in situations where costs cannot be precisely determined, it assumes the more costly of reasonable possibilities. By making the additional needed financial resources available on an annual basis and assuming that these resources were utilized for the described activities, it may be assumed that the conservation objectives will be achieved.

Table 5.14. Additional efforts needed to increase the degree of achievement of the conservation objectives

Additional efforts	Impact of additional efforts on the objective of:		
	Freezing	Accessibility	Adaptation
Increasing numbers of crop varieties in ex situ collections and improvement of the security of ex situ stored accessions, including activities like:			
surveys and inventories	X	X	
collection of PGRFA	X	X	
sustaining existing ex situ collections	X		
regeneration of threatened ex situ accessions	X		
Improvement of the availability of accessions, including activities like:			
expanding the characterization and evaluation activities		X	
developing and improving the information and early warning systems	X	X	
Control of genetic diversity in farmers' fields, including activities like:			
on-farm management and improvement of PGRFA			X
in situ conservation of wild crop relatives and wild plants for food production			X
reintroduction of PGRFA into agricultural systems	X		X
Supporting activities, including activities like:			
supporting or establishing institutions (national programs, networks)	X	X	X
capacity building (education and training, public awareness)	X	X	X

Source: modified according to FAO, 1996b

Table 5.15. Cost estimates for the essential activities to increase the effectiveness of the global conservation efforts

Additional efforts	Approaches annually [US $ '000,000]		
	Basic approach	Moderate approach	High-Urgency Approach
Increasing numbers of crop varieties in ex situ collections and improvement of the security of ex situ stored accessions, including activities like:	38	59	97
surveys and inventories	2	3	8
collection of PGRFA	1	2	3
sustaining existing ex situ collections	30	47	74
regeneration of threatened ex situ accessions	5	7	12
Improvement of the availability of accessions, including activities like:	13	23	43
expanding the characterization and evaluation activities	3	7	15
developing and improving the information management	8	13	22
developing early warning systems	2	3	6
Control of genetic diversity in farmers' fields, including activities like:	18	32	61
on-farm management and improvement of PGRFA	7	17	37
in situ conservation of wild crop relatives and wild plants for food production	5	8	15
reintroduction of PGRFA into agricultural systems	6	7	9
Supporting activities, including activities like:	34	62	120
supporting or establishing institutions (national programs, networks)	13	24	45
capacity building (education and training, public awareness)	21	38	75
Total:	109	185	337

Source: adapted and newly calculated according to Virchow, 1996a

5.4 Effectiveness of PGRFA Conservation for Analyzed Countries

After analyzing the expenditures for PGRFA conservation in 39 different countries, the output of all national conservation activities must be analyzed. Because of the problems associated with the valuation and monetization of the benefits of PGRFA conservation, i.e., the intergenerative value as well as the non-tradable value of PGRFA, a cost-benefit analysis will not be carried out to identify profitable conservation investments. Consequently, the evaluation of the effectiveness of national conservation activities has to be taken as a substitute for evaluating these activities.

Fig. 5.8. Operational indicators determining the assessment of PGRFA conservation

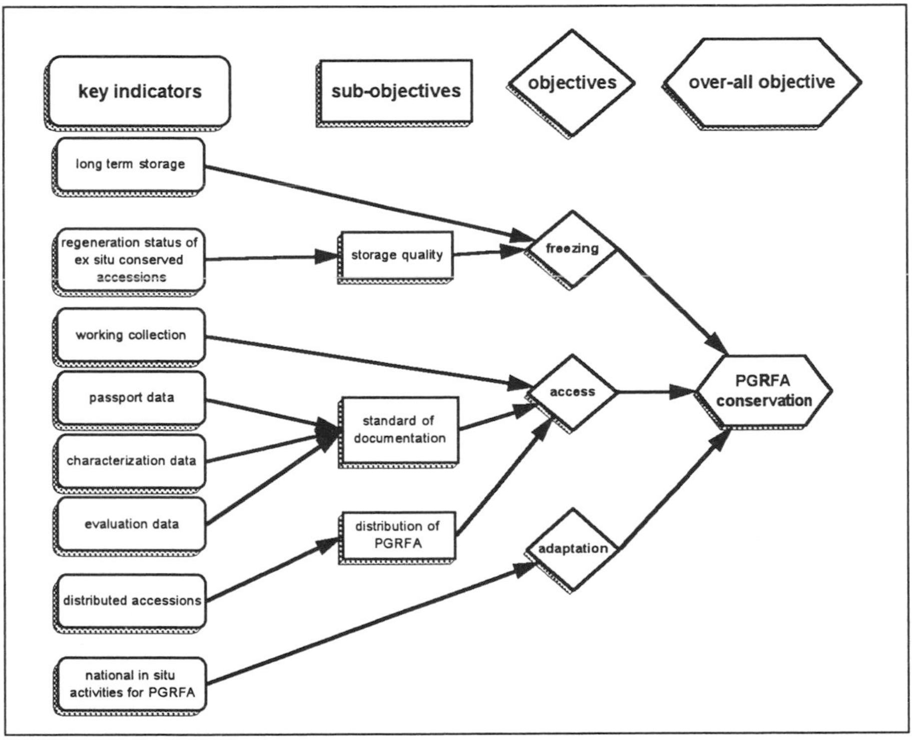

The overall objective is to obtain an idea of successful investment in conservation of PGRFA in different countries. To do so, it is necessary to compare the conservation expenditures of the surveyed countries with the benefits of PGRFA conservation. The conservation success can be identified by judging the degree of goal achievement of the national conservation activities. According to this

approach, the three identified objectives of PGRFA conservation, i.e., freezing for future utilization, accessibility of stored PGRFA, and adaptation to changing environmental conditions, have to be divided into more operational sub-objectives and quantifiable indicators. Because of the general difficulties with the initial stage of conservation activities, the data availability restricts the assessment to a few key indicators (see Fig. 5.8.). Not all surveyed countries could be included in the analysis of their effectiveness in PGRFA conservation[70].
The effectiveness of PGRFA conservation is hereby defined as the degree of goal achievement of the national conservation activities. The aggregate results of the analysis on the effectiveness of PGRFA conservation in the different countries can be seen in Table 5.16. This is based on the disaggregated assessment described below. At the aggregate level it shows that only 15% of 34 countries (5 countries) may be classified as countries with a high total effectiveness. Only 2 countries (6%) were classified as having a medium total effectiveness in their conservation efforts. 27 countries, representing 79% of the surveyed countries, had a low overall effectiveness. It is interesting to note that effectiveness of PGRFA conservation does not necessarily relate to income level of the countries. The group with a high effectiveness of conservation efforts consists of high-income countries like the US, Japan, and Germany, but low-income countries like Ethiopia and countries in transition like Poland also belong to this group. On the other hand, countries like Switzerland, Madagascar, France, and India belong to the group representing countries with a low effectiveness in PGRFA conservation efforts.

Operational sub-objectives and indicators for the first objective - freezing for future utilization, can be defined as the existence of long-term storage facilities on the one hand and as the storage quality on the other hand (see Fig. 5.8).

---

[70] For instance, Lesotho does not have a genebank at present, but plans are being made to build one. Some material has been collected by international institutions during 1989-1991, which is stored in the UK for the time being. Hence, at present no characterization or evaluation has been carried out in Lesotho (CRL, 1995). Therefore, Lesotho was not considered in the analysis. Other countries, like Norway and Finland, belong to the Scandinavian group which is running a regional genebank in Sweden, the Nordic Gene Bank, in which the collections of all Scandinavian countries are stored. Therefore, the single countries do not conserve the material in their own country, but rather keep some working collections stored. Hence, Norway's and Finland's information would distort the survey. Therefore, they were excluded from the effectiveness analysis as well. Even though Italy provided information on their expenditures related to PGRFA conservation, there was not enough substantial information to involve Italy in this analysis. Therewith, 34 countries were included in the survey.
Countries included in the survey are: Austria, Belarus, Brazil, Canada, China, Cyprus, Czech Republic, Egypt, Ethiopia, France, Germany, Greece, Haiti, India, Ireland, Japan, Madagascar, Pakistan, Peru, Poland, Portugal, Romania, Russia, Seychelles, Slovak Republic, South Africa, Spain, Suriname, Switzerland, Tanzania, Togo, Tonga, the United Kingdom, and theUnited States of America.

Table 5.16. Rate of total effectiveness of PGRFA conservation in different countries

Rate of total effectiveness of PGRFA conservation	Classified countries	% of surveyed countries
High	Ethiopia, Germany, Japan, Poland, USA	15%
Medium	Czech Republic, Canada	6%
Low	Austria, Belarus, Brazil, China, Cyprus, Egypt, France, Greece, Haiti, India, Ireland, Madagascar, Pakistan, Peru, Portugal, Romania, Russia, Seychelles, Slovak Republic, South Africa, Spain, Suriname, Switzerland, Tanzania, Togo, Tonga, United Kingdom	79%

The rate of total effectiveness of PGRFA conservation is defined as the degree of goal achievement of the national conservation activities. See Table 5.19 for the detailed rating.

Long-term storage facilities: Even though it is possible to freeze genetic resources in medium and short-term storage facilities, a safe and high quality freezing potential for future utilization is significantly decreasing from long-term storage over medium-term storage to short-term storage. Long-term storage facilities have to meet international standards. The facilities must have a reliable power supply, and procedures for safe duplication and regeneration have to be carried out (FAO/IPGRI, 1994). The long-term storage facilities are listed in Table 5.19; 68% of the countries surveyed have long-term facilities for the conservation of PGRFA, 11 countries do not have long-term facilities for the conservation of PGRFA germplasm. Countries without long-term storage facilities may be in need of some, but because of the lack of resources, financial and human capital, these facilities have not been established in these countries. On the other hand, a country like Russia - with one of the world's largest genebanks - has no long-term storage facilities. Its conservation policies differ in the basics of conservation management. Russia, or better the Vavilov Institute since the time of the Soviet Union, has been storing germplasm on a more medium and short-term base, utilizing the instruments of frequent regeneration and collections as major tools in their conservation efforts. But since the political transition, the Vavilov Institute has had difficulties with its regeneration and collection activities, because the places of origin are no longer in the same country or in the same political sphere of influence, creating institutional and political problems (FAO, 1996h). Consequently, the quality of the Russian long-term storage activities can be assessed to be poor at present.

Storage Quality: The existence of long-term storage facilities is an essential, but not a sufficient condition for a safe and high quality long-term conservation of genetic resources. Hence, the storage quality is an additional sub-objective to be incorporated into the evaluation. The present situation in the ex situ conservation facilities concerning the accessions to be regenerated was chosen as an operational key indicator (see Fig. 5.8). Several other indicators could be utilized to indicate the quality standard of the storage facilities, e.g., health indicators, loss of genetic resources etc. But because of limited information available, the accessions still to be regenerated present the most useful information, because regeneration of any reproductive plant material in storage is an important part of the work of any genebank and characterize the general conditions of a genebank (FAO/IPGRI, 1994). If the viability falls below 85% of its initial value, the accession is threatened by extinction (Holden and Williams, 1984). Whereas some genotypes may loose their viability quicker than others, it is a general rule that regeneration take place every 10 years (see Chap. 2.4) in order to guarantee the viability of the accessions. By taking the average over a time horizon of 10 years, a genebank must regenerate approximately 10% of its stored germplasm every year, to meet this standard in the longterm. Consequently, the quality standard of the analyzed national ex situ conservation facilities has been divided into three categories:

- good quality standard, if less than 20% of the facility's accessions were in need of regeneration;
- basic quality standard, if more than 20%, but less than 50% of the facility's accessions were in need of regeneration;
- poor quality standard, if less than 50% of the facility's accessions were in need of regeneration.

As can be seen in Table 5.19, only 8 (24%) of all analyzed countries could be classified as having a good quality standard of long-term facilities. Additionally 14 countries (41%) have basic quality standards, whereby still 12 countries (35%) have poor quality standards; this includes those countries without any long-term storage facilities. It is worth highlighting that only Japan, Ethiopia and Poland report that less than 10% of total genebank accessions require regeneration (Country Reports, 1996). Because of the lack of information, some assumptions had to be made concerning the state of regeneration in some storage facilities (see Table 5.19 and Appendix 15).

After analyzing the sub-objectives and key indicator, the degree of goal achievement for the objective "freezing" can be assessed by combining the quality standard of the storage facilities with the existence of long-term storage facilities according to the following matrix (Table 5-17). The results can be seen in Table 5.19. The key question for the first objective was whether a country's conservation activities may guarantee the long-term conservation of PGRFA - highlighting the conservation of genetic resources for future use. It has been shown that even though all analyzed countries have conservation activities, over 50% of these countries have poor qualities regarding the long-term storage of PGRFA.

The key question for the second objective relates to the present use of PGRFA. The demand for PGRFA is determined mainly by the conventional breeders at present, but the demand by the biotechnology industry is increasing. The more the demand specializes in specific genetically coded functions and even more detailed to specific genetically coded information, the more important will be the accessibility of genetic resources.

Table 5.17: Assessment of the degree of goal achievement for "freezing"

Long-term storage	Quality standard of storage		
	Good:	Basic:	Poor:
Existent:	Good	Basic	Poor
Non-existent:	-	Poor	Poor

The accessibility will depend upon the physical access to the germplasm as well as on its state of processing. Therefore, the degree of goal achievement for the second conservation objective, i.e., the quality standard of the accessibility of PGRFA in a country, can be divided into 3 sub-objectives: (1) the existence of working collections, (2) the documentation of stored accessions, and (3) the number of accessions distributed. The first two mentioned sub-objectives describe the potential accessibility of germplasm for users, whereas the third sub-objective describes the actual accessibility of germplasm in 1994.

Users of PGRFA will try to minimize their search costs by looking for specific genetically coded functions or genetically coded information. Search costs are determined by allocating resources to find the specific information necessary. Initially, search costs are incurred by the physical movement to farmers' fields (expenses for travel, utilized material and opportunity costs for travelling) to look for specific traits in specific crops. The search costs can be reduced significantly by storing genetic resources in more centralized storage facilities than farmers' fields, which means an increase of the accessibility of existing national genetic resources. Therefore, a first indicator for the degree of accessibility of genetic resources in a country could be the coverage of existing diversity in the country by national storage facilities. But at present, there is not enough information to analyze how representative current ex situ collections are of total national diversity, because of a lack of a comprehensive inventory of PGRFA. Over 50% of the analyzed countries have stressed the lack of knowledge of existing indigenous plant genetic resources[71] and the other countries did not mention the degree of coverage at all.

Working collection: To measure the degree of goal achievement for the accessibility of conserved genetic resources, the physical presence of the germplasm has to be ensured. Long-term storage facilities are the best mechanism to conserve genetic resources for future needs, but these facilities are not very flexible in regard to the present demand. It takes time to cautiously unfreeze the germplasm and it has a negative impact on the viability of the germplasm each time the accession is taken out of the cold chamber. A working or active collection is needed to provide germplasm to demanding users and still keep a good quality standard for the germplasm conservation. The working collection may be used for

[71] Austria, Brazil, China, Cyprus, Egypt, Ethiopia, Germany, Haiti, India, Ireland, Japan, Norway, Poland, South Africa, Spain, Tanzania, Togo, the United States of America.

the documentation work as well as for the exchange of accessions. Consequently, a conservation system with long-term storage facilities, but without working collections significantly reduces the potential of accessibility of genetic resources for present use. Hence, the existence of a working collection is the first indicator for a good accessibility to conserved genetic resources.

As can be seen in Table 5.19, all relevant countries, except Haiti, Seychelles, and Suriname, have working collections. Most of the analyzed countries have the potential to provide germplasm to whoever requests it.

Documentation: In addition to the physical existence and the flexible handling of germplasm practiced with working collections, the quality standard of the documentation of stored germplasm indicates the degree of accessibility to PGRFA. The documentation of stored germplasm is a very important component for the users of PGRFA in their decision making and research process. Better information result in less search costs. One main factor of the search costs, apart from the costs for looking for the physical existence, is the time spent in producing the necessary information, if it is not available. The user is able to collect most of the necessary data only by cultivating the germplasm. Not only the information quantity, but also the information availability determines the accessibility to genetic resources. Even though accessions may be well documented in a specific storage facility, as long as this information is not accessible for all the potential users, for instance through Internet, the actual exchange of specific accessions will not occur, even though there is a demand for it. Consequently, with increasing information content and increasing information availability, the accessibility to PGRFA is increasing as well.

A good quality standard in the germplasm documentation is necessary for the accessibility of the accessions. The more data is available, the higher will be the specific information value. Hence, the percentage of available passport, characterization, and evaluation data for the germplasm collection may be used as key indicators in determining the quality standard of the documentation. In this context, good quality standards for the documentation are defined by the existence of more than 80% passportdata and more than 50% other data (characterization and evaluation data) in the whole collection. Basic quality standards must have more than 80% passport data and 30 to 50% other data. If a collection has less than 80% passport data and less than 50% other data or even 100% passport but no other data, then the documentation of the collection is considered having a poor quality.

It must be stressed once again: although international standards for PGRFA conservation exist, the amount and quality of information included in the passport data of accessions in many collections may, however, be minimal or different. Some passport data only contain information about country of origin (Peeters and Williams, 1984).

As could be seen while assessing the goal achievement for the first objective, i.e., freezing, these measurements chosen are rough estimates. Therefore, other indicators determining the accessibility of accessions more precisely, e.g., the existence of more user-friendly documentation systems with standard formats or computerized data available on the Internet, are not included due to the present lack of necessary information from the majority of countries. The complexity of the problem is increased because the documentation standard in some countries may be better than judged in this process. Because of a decentralized national conservation policy, the documentation data are not always distributed well. Consequently, they do not appear in the Country Report. The negative externalities of an uncoordinated decentralized conservation and information management are the reduced accessibility of genetic resources for the various users. These users may search for specific genetically coded information or functions by utilizing the information accessible on national level and consequently fail to find the existing information, which is not to be found on national level. The transaction costs for searching for genetically coded information will be high in a decentralized conservation system without a well functioning information management. Hence the potential users of the specific genetically coded information will either find the same or similar information in another country or will change the direction of research. Consequently, the accessibility of germplasm stored in ex situ storage facilities will decrease, based on the degree of national decentralization of PGRFA conservation combined with a lacking or bad information management. The Country Reports' information on the national documentation standard as an indicator reflects this situation.

According to the rough estimation utilized, 56% of the analyzed countries have poorly documented ex situ storage facilities (see Table 5.19 and for more detailed information Appendix 16) and only 25% of the countries show a good quality standard in their documentation of ex situ stored PGRFA. This assessment is determined mainly by no or only very little other data available except passport data.

Distribution of accessions: The third and final key indicator for the degree of goal achievement of the second objective is the distribution of accessions in 1994. An important role of conservation facilities is to promote and facilitate the distribution of their stored germplasm. This involves international movement of germplasm as well as the distribution at the national level - usually to plant breeders and other researchers, but only seldom directly to farmers. In general, germplasm has been freely available to bona fide users upon request. Consequently, annual germplasm distribution reflects best the actual state of accessibility of the conservation facilities, also this may be partly an issue of lack of effective demand in the country concerned. If there had been sufficient data for every surveyed country, germplasm distribution could be the only key indicator needed for assessing the goal achievement of accessibility. Only 10 countries, however, reported the germplasm movement in their storage facilities (see Table 5.19 and for more detailed information Appendix 17). Hence, this indicator was used in addition to the other two indicators discussed above. If the distribution of germplasm in 1994 was 10% of the stored accessions or more, the standard of distribution was judged as good quality standard. If the conservation facilities distributed by less than 10% but more than 3% of the stored accessions, they received a basic quality standard. If the conservation facilities distributed less than 3% of their stored material, their quality standard is defined as poor. Only three countries have a good quality standard of germplasm distribution: the USA which distributed approximately 128,000 accessions in 1994 (23% of their stored material), Ethiopia with 6,000 (11%), and Germany with 19,000 (10%). It is interesting to note the existence of countries with a high amount of stored germplasm but with only very little germplasm movement. Russia (with over 300,000 accessions stored) and Brazil (over 190,000) have distributed less than 1% of their germplasm in the year 1994. This substantiates the impression made through the information provided by the Country Reports and the discussions in the past years, that genebanks are still seen and managed mainly as if they were facilities only to store the diversity of PGRFA but without any or only a little connection to the utilization of the stored material. Even a country like Japan, with well-equipped storage facilities and one of the largest genebanks worldwide (over 200,000 accessions) has distributed less than 7,000 accessions in 1994 (which amounts to 3% of their stored accessions). (Country Reports, 1995).

The goal achievement for the accessibility of stored accessions is determined by the quality of the above discussed key indicators and sub-objectives. By utilizing the matrix described in Appendix 18, an overall quality standard for the accessibility can be defined for all the relevant countries (see Table 5.19 for the results).

The most important result is that only 6 countries have a good access standard of their conserved accessions - making up less than 17% of all surveyed countries

National in situ activities: Because of the lack of sufficient information, the existence of in situ conservation activities was taken as the only available indicator to assess the goal achievement of the third overall objective - the adaptation of PGRFA to changing environmental conditions. As stated earlier (see Chap. 2.4), in situ conservation activities in the field of conservation of PGRFA diversity is still uncommon. Therefore, it is not surprising that only 30% of the surveyed countries mentioned in situ conservation activities (see Table 5.19). There are probably some more in situ activities in the relevant countries, but they were not mentioned in the Country Reports, reflecting that these not-mentioned activities are supported by NGOs and are not linked to national activities.

Finally, the total effectiveness of PGRFA conservation according to the objectives defined can be obtained by merging the results of the three partial effectiveness' analyses (see Table 5.19). The same approach was undertaken as it was utilized to define the quality of the accessibility. A matrix was designed with all three objectives and their different levels of effectiveness to determine the total effectiveness of the conservation efforts in the countries (see Table 5-18), based on the calculations made in Appendix 19).

Table 5.18. Overall quality for PGRFA conservation determined by the conservation objectives

Freezing		Access: Poor	Access: Basic	Access: Good	Adaptation
Freezing	Good			Ethiopia, Japan, USA	Existent
		Austria, China, Switzerland		Poland	Non-existent
	Basic		Canada	Germany	Existent
		Ireland, Portugal, Romania	France, Greece, Pakistan, UK	Czech Rep.	Non-existent
	Poor	Egypt, Haiti, India, Surname	Peru		Existent
		Belarus, Madagascar, Russia, Seychelles, Slovak Rep., Tanzania, Togo, Tonga	Brazil, Cyprus, South Africa, Spain		Non-existent
		Poor	Basic	Good	
		Access			

whereby the overall quality is illustrated by:

low:	(dark shading)
medium:	(light shading)
high:	(no shading)

The effectiveness of the countries' conservation activities may be assessed and ranked in accordance to the outline above. Table 5.19 summarizes the assessment for the single objectives as well as the rating for the total effectiveness.

Summarizing this chapter, it can be highlighted that approximately US $ 800 million are annually invested in the conservation management of PGRFA. These expenditures include ex situ conservation, in situ conservation as well as the costs for the institutional process of developing a conservation and exchange system. Another important cost factor in the conservation management is the information processing, which is of essential importance for the future role of storage facilities.

Global annual unit costs for ex situ conserved accessions were calculated based on different cost data. The costs amount to an average of US $ 44. This seems acceptable in the light of optimizing the different needs of long-term freezing for future needs and the short-term or medium-term accessibility of the germplasm for breeders' utilization. Different conservation techniques may, however, reduce the ex situ conservation costs. Still in the early stage of introduction, cryopreservation may take over the major role in conserving germplasm for the long-term future without any accessibility for medium- or short-term access. Already, cryopreservation is presently the cheapest conservation technique, with an average of US $ 20 (see Table 5.7). If this foreseeable trend takes place, the need for other long-term storage facilities will decrease significantly, reducing the financial resources necessary for the establishment and maintenance.

Table 5.19. Effectiveness of conservation of PGRFA in national programs

	Partial Effectiveness of the Objectives of Conservation and Utilization PGRFA in National Programs										Total Effectiveness of PGRFA
	Freezing			Access				Adaptation			
	Long-term storage	Quality standard of storage	Freezing	Working collection	Standard of documentation	Standard of distribution of PGRFA[72]	Access	National in situ PGRFA activities	Adaptation		
Austria	e.	good	good	e.	poor	n.i.	poor	n-e.	n-e.		low
Belarus	n-e.	basic	poor	e.	poor	n.i.	poor	n-e.	n-e.		low
Brazil	e.	poor	poor	e.	basic	poor	basic	n-e.	n-e.		low
Canada	e.	good	good	e.	basic	basic	basic	n-e.	n-e.		medium
China	e.	good	good	e.	poor	n.i.	poor	n-e.	n-e.		low
Cyprus	n-e.	basic	poor	e.	basic	n.i.	basic	n-e.	n-e.		low
Czech Republic	e.	basic	basic	e.	good	basic	good	n-e.	n-e.		medium
Egypt	e.	poor	poor	e.	poor	n.i.	poor	e.	e.		low
Ethiopia	e.	good	good	e.	good	good	good	e.	e.		high
France	e.	basic	basic	e.	good	n.i.	basic	n-e.	n-e.		low
Germany	e.	basic	basic	e.	good	basic	good	e.	e.		high
Greece	e.	basic	basic	e.	basic	n.i.	basic	n-e.	n-e.		low
Haiti	n-e.	poor	poor	n-e.	n-e.	n.i.	poor	e.	e.		low
India	e.	poor	poor	e.	poor	n.i.	poor	e.	e.		low
Ireland	e.	basic	basic	e.	poor	n.i.	poor	n-e.	n-e.		low
Japan	e.	good	good	e.	good	basic	good	e.	e.		high
Madagascar	n-e.	poor	poor	e.	poor	n.i.	poor	n-e.	n-e.		low
Pakistan	e.	basic	basic	e.	basic	basic	basic	n-e.	n-e.		low
Peru	n-e.	basic	poor	e.	basic	n.i.	basic	e.	e.		low
Poland	e.	good	good	e.	good	basic	good	n-e.	n-e.		high
Portugal	e.	basic	basic	e.	poor	n.i.	poor	n-e.	n-e.		low
Romania	n-e.	basic	poor	e.	poor	n.i.	poor	n-e.	n-e.		low
Russia	n-e.	basic	poor	e.	poor	poor	poor	n-e.	n-e.		low
Seychelles	n-e.	poor	poor	n-e.	poor	n.i.	poor	n-e.	n-e.		low
Slovak Republic	n-e.	basic	poor	e.	poor	n.i.	poor	n-e.	n-e.		low

[72] Data of PGRFA distribution are from 1994.

Table 19 contd.: Effectiveness of conservation of PGRFA in national programs

	Partial Effectiveness of the Objectives of Conservation and Utilization PGRFA in National Programs									Total Effectiveness of PGRFA
	Freezing			Access				Adaptation		
	Long-term storage	Quality standard of storage	Freezing	Working collection	Standard of documentation	Standard of distribution of PGRFA[73]	Access	National in situ PGRFA activities	Adaptation	
South Africa	e.	poor	poor	e.	basic	n.i.	basic	n-e.	n-e.	low
Spain	e.	poor	poor	e.	basic	n.i.	basic	n-e.	n-e.	low
Suriname	n-e.	poor	poor	n-e.	n-e.	n.i.	poor	e.	e.	low
Switzerland	e.	good	good	e.	poor	n.i.	poor	n-e.	n-e.	low
Tanzania	e.	poor	poor	e.	poor	n.i.	poor	n-e.	n-e.	low
Togo	n-e.	poor	poor	e.	poor	n.i.	poor	n-e.	n-e.	low
Tonga	e.	poor	poor	e.	poor	n.i.	poor	n-e.	n-e.	low
UK	e.	basic	basic	e.	good	n.i.	basic	n-e.	n-e.	low
USA	e.	good	good	e.	good	good	good	e.	e.	high

exs.: existent; n-exs.: non-existent

Furthermore, it may be of economic advantage to differentiate the conservation approach by implementing a global conservation system because of economies of scales. This may be based on one "mega-genebank", holding all of the approximately 3 million distinct accessions of PGRFA[74]. Duplicates could be stored in two other genebanks[75]. Perma frost should be utilized as the conservation technique, reducing the unit costs considerably [76]. Hence, at a regional or national level genebanks with medium-term storage capacities could function as

[73] Data of PGRFA distribution are from 1994.

[74] Even though there are around 6.2 million accessions conserved in genebanks worldwide, FAO - through the country-driven process - estimates that there are only between 1 and 2 million unique accessions (FAO, 1996a). If we estimate that an average 70% of PGRFA germplasm has been already collected, the capacity for 3 million accessions is sufficient. But even increasing the capacity to 4 million would not be a high investment.

[75] Already today 2 safety duplicates are commonly accepted to represent enough safety for the long-term conservation (FAO, IPGRI, 1994).

[76] The genebanks, established and run in international responsibility, are build in cold areas of the world and utilized as perma-frost conservation facilities (e.g., North Scandinavia, Switzerland, India or some other place with - 20 °C). They have to be equipped with a good information and service unit. For more detail see Virchow, 1997a.

coordinators between the long-term storage and the short-term needs of breeders and other researchers.

On the one hand, in situ conservation of PGRFA, understood as maintaining landraces in farmers' fields, is undertaken by farmers as a positive external effect. Hence, no costs are attributed to this approach of in situ conservation. On the other hand, programs and projects are conceptualized to promote in situ conservation. Until now, few of these have been economically analyzed. Of the known results, it seems, however, that high costs may be involved with this approach. One example in Peru indicates program costs of US $ 500 per accession and year. Even if it may be assumed to be a pilot project including breeding components, transferable to other regions for less costs, it cannot be economically sound and sustainable to promote similar projects.

The analysis of the effectiveness of storage facilities in 34 countries suggests that the present costs for the conservation of PGRFA at the national and international level do not give much evidence concerning the impact of conservation activities. Some countries with very different expenditure structures have received a high rating by achieving all three conservation objectives. The result implies that existing financial resources should be used more effectively. The emphasis must, therefore, be on measures which improve the efficiency of conservation through the rationalization of efforts, e.g., the rationalization of collections through regional and international collaboration, an improved data and information management, as well as the reduced over-duplication of samples. Furthermore, mechanisms should allow countries to place materials in secure storage facilities outside their borders, without compromising their sovereign rights over such material.

6 Two Examples of PGRFA Conservation Systems: India and Germany

In Chapter 5.2 the different countries involved in PGRFA conservation were characterized according to their expenditures and state of agrobiodiversity. There are the supply-driven spenders on the one side – countries identified by high expenditures for the conservation, a high level in agrobiodiversity as well as the technological potential for national breeding activities. The countries' objectives are to promote the national breeding efforts as well as to be prepared for a future free exchange of PGRFA. India is one of the countries in this category.

Germany, however, as a demand-driven spender, is interested in safeguarding the breeding industry's demand for genetic resources. The demand driven spenders are agrobiodiversity-poor countries with a high-standard breeding sector. These countries invest in the maintenance of their ex situ collections and information processing. Furthermore they are interested in a secure in situ conservation system, especially in countries rich in agrobiodiversity. The demand-driven spenders are willing to compensate the in situ conservation in agrobiodiverse countries to protect PGRFA as future input for breeding.

6.1 Indian Case Study[77]

India covers 3,287,263 km^2, of which 60% is used for agricultural production. India, representing only 2.4% of the total world's area, has a population of more than 920 million people, which makes up 16% of the world's population. India is very rich in plant biodiversity, with 20,000 vascular plant species, of which approximately 7,000 are endemic (ICR, 1995). The 20 agro-climatic zones of the country represent all of the existing agro-climatic zones, ranging from arid to humid ecosystems and from highland to island ecosystems (Chandel, 1995). In all these ecosystems, 160 crop species are cultivated, of which 35 species were domesticated in the Indian gene center (Chandel and Rana, 1995). Crops like kodo millet, aubergine, mango, black pepper, and some types of rice originate in the gene center of the Hindustani area (Zeven and Zhukovsky, 1975). The major

[77] Important information was collected during the visit to India in 1996 and through the Indian Country Report submitted to FAO.

diversity can still be found in rice, with estimated 16,000 landraces in farmers' fields (ICR, 1995).

6.1.1 Indian National Plant Genetic Resources System

The focal point of PGRFA conservation and management in India is the Indian National Plant Genetic Resources System (INPGRS), operated by the National Bureau of Plant Genetic Resources (NBPGR), and established in 1976 (Fig. 6.1.). NBPGR is an independent organization within ICAR (Indian Council of Agricultural Research), accountable to the Department of Agricultural Research and Education (DARE) of the Union Ministry of Agriculture. The bureau is staffed with approximately 550 employees, of whom approximately 150 are scientists (Chandel, 1996).

Fig. 6.1. Indian National Plant Genetic Resources System

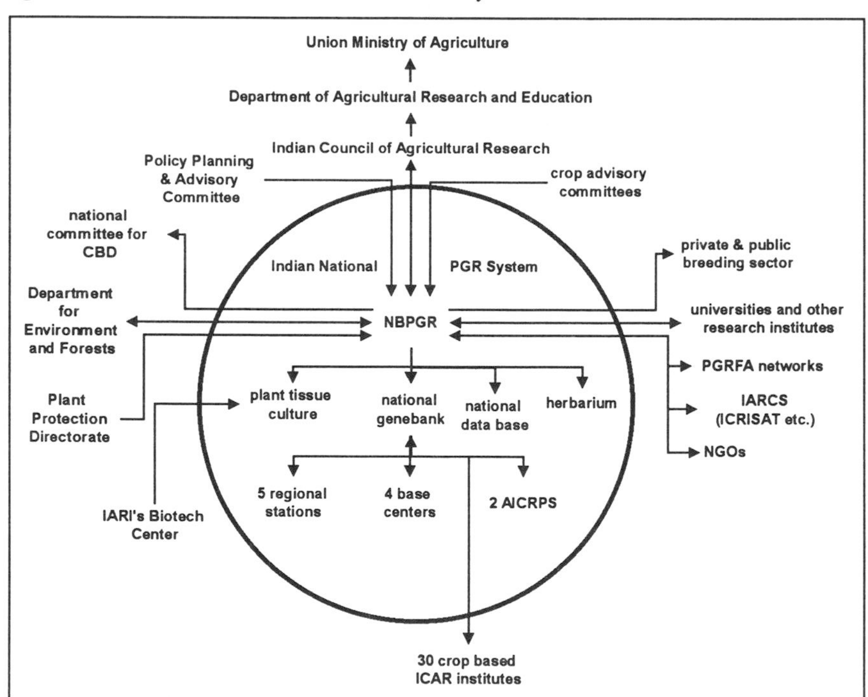

NBPGR's mandate includes planning, organizing, conducting, promoting, coordinating and leading all activities concerning PGRFA ex situ conservation and management. This includes activities like basic research and training programs at all national and international levels. Furthermore, NBPGR is responsible for the health control of all PGRFA, imported and exported for research purposes. The quarantine powers and authority are delegated to the Director of NBPGR from the

Plant Protection Directorate, Government of India. From the start, emphasis was laid on the collection and conservation of crop plants from within the country as well as the improvement of existing and utilized crops (Paroda, 1986). Conservation is seen as means for immediate or near-distant utilization of genetic resources in breeding; therefore, the need for an effective evaluation has been identified (Arunachalam, 1986).

The main components of INPGRS are the PGRFA storage facilities, namely the national genebank, the national facility for plant tissue culture (NFPTCR), and the national seed museum and herbarium of cultivated plants and their wild relatives. The genebank and the tissue culture facility hold base collections together approximately 144,000 accessions of the different agricultural crops (see Fig. 6.2). A further 197,999 accessions are stored in working collections (ICR, 1995), making a total of 342,108 accessions stored in India. The rate of duplicates is not known, however.

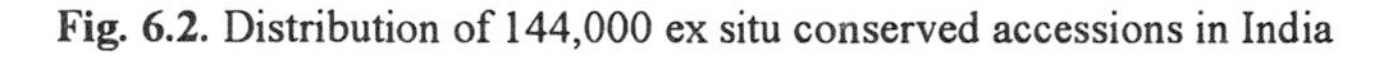

Fig. 6.2. Distribution of 144,000 ex situ conserved accessions in India

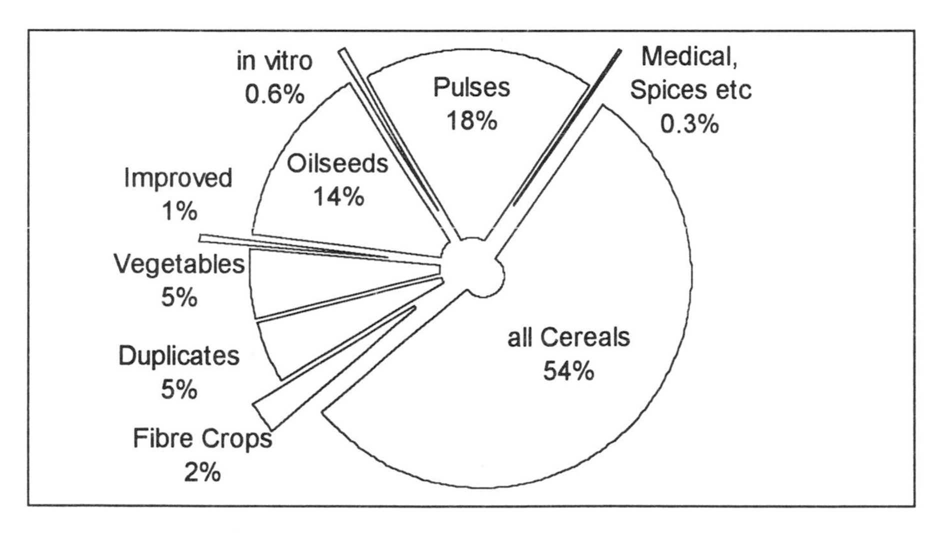

Note: Except 849 accessions, which are stored in in vitro conservation facilities, all other accessions are conserved ex situ in genebanks.
Source: recalculated according to data from ICR, 1995

As a long-term conservation facility, the national genebank holds some 143,000 accessions and has a capacity for 250,000 accessions. An additional conservation facility, financially supported by the USA, was inaugurated in November 1995. This facility has an additional storage capacity for approximately 1 million accessions and the storage facilities for approximately 250,000 samples in cryopreservation (Paroda, 1996). The facility is backed with all of the necessary conservation and processing equipment.

NBPGR activities in the ex situ conservation are supported by its own Regional Stations Network. These five regional stations are located in different agro-climatic zones (ICR, 1995) and are supplemented by an additional 4 base centers and some 30 'National Active Germplasm Sites', as crop-based ICAR institutes (Paroda, 1986). These organizations extend the influence and coordination of NBPGR into the different Indian states and enable NBPGR to meet its obligations in the different agro-ecological regions of India. Further cooperation exists with two All-India Coordinated Research Projects (AICRPS), which are located at NBPGR and focus on specific crops.

On the other hand, the Policy Planning and Advisory Committee may be seen as the main advisory board for PGRFA conservation and management by NBPGR. It is constituted by the ICAR and provides policy guidelines and oversees the implementation of the activities (Paroda, 1996). Furthermore, the Crop Advisory Committees guide the activities of NBPGR through review and advise for a single crop or for a group of crops (ICR, 1995).

NBPGR is funded by ICAR. The expenditures for the financial year 1994/95 amount to a total of US $ 2 million (see Table 6.1). This includes the expenditures directly related to the work of NBPGR amounting to US $ 1.2 million as well as to the Indian share in connection with foreign cooperation of US $ 1 million. The foreign-aided programs at NBPGR are the assistance in strengthening and upgrading the conservation facilities, joint germplasm explorations, joint research and training programs on the one side (USAID and Indio-UK project), and two smaller projects, concentrated on the collection of native diversity and the strengthening of the All India Coordinated Crop Improvement Programs on the other hand (the national seed project). The overall total of the Indian domestic expenditures for the national conservation program amounts to US $ 6.8 million, which is comparatively high for a developing country (see Chap. 5.2).

In addition to the ex situ conservation activities, there are some ongoing in situ conservation activities. The Department of Environment and Forests, Govt. of India is responsible for the implementation of programs related to biodiversity conservation and for the negotiations of CBD. Hence, the Department of Environment and Forests is responsible for all in situ conservation activities in India, in particular the establishment of "Biosphere Reserves". NBPGR assists the department in the establishment of these projects and, as a member of the national committee, the Department takes an active role at the national level for the implementation of CBD (Chandel and Rana, 1995).

NBPGR's documentation system is set up as the National Data Base serving in all relevant collaborations. The national collaborations are with the Department of Environment and Forest and with the Department of Biotechnology, Government of India, the Indian Council of Forestry Research and Education, and with Botanical Survey of India. The Department of Biotechnology has established the plant tissue culture facility at NBPGR. IARA's Biotechnology Center is closely linked to NBPGR by assisting the NPTCR and the National Center on DNA Fingerprinting and Gene Sequencing. Furthermore, links to various agricultural universities exist as well as to the crop-based institutes of ICAR. In addition to this cooperation, NBPGR assists 13 other collaborating institutes and centers in establishing storage facilities. NBPGR is also in touch with NGOs active in PGRFA conservation, for instance the M.S. Swaminathan Research Foundation - a nonprofit, political organization established in 1988. The foundation's work is focused on environmentally sustainable and socially equitable development. The foundation actively influences the Indian policy discussion in the area of PGRFA conservation and has created a community genebank to store collected seeds (Swaminathan, 1996d). The Honey Bee Network has the objectives to strengthen the links between formal and informal knowledge systems. One major output is the database on local innovations and traditional ecological and technological knowledge. The organization is supported by the Society for Research and Initiatives for Sustainable Technologies and Institutions as a NGO, and by the Indian Institute of Management (IIMA) (Gupta, 1996).

Table 6.1. Total expenditures for the Indian national program for the conservation of PGRFA in 1994/95

	Domestic expenditures in US $	Foreign assistance in US $	Total expenditures in US $
NBPGR	1,239,374		1,239,374
National Seed Project II (World Bank funded)		22,514	22,514
Collection and conservation of temperate fruits (US funded)		7,625	7,625
Indo-USAID PGR Project			
Indian share	923,029		923,029
US share		2,007,778	2,007,778
Indo-UK PGR Project			
Indian share	48,398		48,398
UK share		2,527,133	2,527,133
Total	2,210,801	4,565,050	6,775,851

Source: compiled according to data from Siddiq, 1996

On an international level, NBPGR's collaborations are with the International Agricultural Research Centers (IARCS), above all with ICRISAT at Hyderabad, but also with other centers. The collaborations are concentrated on joint collection missions, germplasm and information exchange, as well as on the training of personnel. In addition to multi-lateral cooperation, some bilateral programs do exist with countries in Europe, USA, Canada, Australia, New Zealand, Russia and others. The cooperation with the USA, for instance, was a 9-year program. It included the construction of the new conservation facility in New Delhi, 4 green houses as well as some training programs, research collaboration, and joint collections in and outside of India (Chandel, 1996).

On the institutional level, India has ratified the Convention on Biological Diversity and is a member of FAO's Commission on Genetic Resources for Food and Agriculture (CGRFA). India adheres to the International Undertaking and strongly supports Farmers' Rights. India is a member of the International Plant Protection Convention. This convention determines the quarantine level and its code of conduct for the import and export of germplasm. India has a seed certification legislation as a seed quality control. Furthermore, India is a member in the plant genetic resources network for South Asia. (FAO, 1996a). India's government is committed to the free exchange of germplasm resources for their active use related to the national and international exchange of PGRFA (ICR, 1996), whereas the Indian policy for microorganism differs significantly. Service charges (amounting to approximately US $ 850) are charged for the identification and testing of soil samples. Furthermore, the export of soil, which must be certified, has been noticeably restricted (Sarbhoy, 1996).

Table 6.2. Exchange of PGRFA by NBPGR, India

Year of introduction	Import to PGRFA	Export from PGRFA
1976	85,872	7,055[b]
1977	74,835	10,686
1978	117,279	8,697
1979	130,194	5,287
1980	51,906	1,917
1981	53,264	2,260
1982	42,663	1,748
1983	49,268	2,683
1984	38,992	3,843
1985	85,117	1,355
1986	52,767	5,535
1987	52,642	2,260
1988	53,629	2,168
1989	50,536	3,310
1990	49,521	1,195
1991[a]	56,735	1,221
1992[a]	52,555	4,674
1993[a]	71,381	3,243
Total over the years	1, 169,156	69,137

Source: recalculated according to Rana and Chandel, 1992; [a]: Chandel and Rana, 1995; and [b]: Paroda and Aroba, 1986

In addition to this specific cooperation, there always has been an intensive exchange of PGRFA between NBPGR and other national and international organizations since its establishment in 1976. As can be seen in Table 6.2, the import of PGRFA from organizations in other countries far exceeds the export of genetic resources. This reflects the intensive Indian breeding activities, especially in the late 1970s, when most of the imported germplasm was used for breeding and only partially conserved. In the beginning, more material was coming in the form of international nurseries; today, however, emphasis is laid on elite material, resistant donors, and wild species (Rana and Chandel, 1992). Furthermore, the figures show that the interest in PGRFA from NBPGR was not very high over the years, or that the export was hampered by restrictive policies. If the distribution rate is calculated based on the amount of exported accessions in 1993 (3,243 accessions) and the present stock of accessions in the national storage facilities (144,000 accessions), the Indian national program has a distribution rate of 2.3%.

6.1.2 National Policy on Plant Variety Recognition in India

The Indian Patent Act of 1970 presently permits only process patents, not, however, product patents in agricultural products and pharmaceuticals. For this reason, India is gaining a bad reputation; USA trade association called India a *"... haven for bulk pharmaceutical manufacturers who pirate the intellectual property of the world's pharmaceutical industry."* (quoted from The Indian Express, 1996).

According to the Article 70.8 of the Trade-Related Intellectual Property Rights (TRIPS) agreement (OECD, 1996a), India must change its IPR system to implement its obligations under the TRIPS until the year 2005.

As regards plant variety protection, India will soon develop an effective suigeneris Plant Variety Protection Act, which will provide a suitable legal framework as required under the provisions of GATT. India is considering adopting the 1978 UPOV Convention, however maintaining the breeder's exemption and, as part of the concept of Farmers' Rights, to entitle farmers to a suitable compensation for their efforts in maintaining agrobiodiversity (Swaminathan, 1996a). The government of India plans to establish a National Community Gene Fund for rewarding farmers' efforts in maintaining agrobiodiversity (Swaminathan, 1996b). It would be the first legal recognition and reward of Farmers' Rights, based on the remuneration rights (Sehgal, 1996). NBPGR is expected to play a vital role in the proposed legislation (ICR, 1995).

There are some major obstacles in the Indian sui generis system. It has been considered to make it compulsory for all breeders to deposit a reference seed sample in the national gene bank and to catalogue it in the National Register, besides the compulsory seed certification and licensing. Especially private breeders fear that their breeding secret will be revealed and their seed may be utilized for further breeding without their agreement (Narayanan, 1996). Furthermore, the Indian patenting procedure takes one to two years at present, which is not an incentive for research and development in India (Senrayan, 1996).

6.1.3 Production and Distribution of Quality Seed

The seed industry, rapidly developing and changing because of institutional changes (Singh et al., 1995), consists of various agencies, which are active in enabling the seed availability of food crops, oilseeds and pulses, horticultural crops, medicinal and aromatic plants, underutilized crops, commercial crops and agro forestry. These agencies are:

- NBPGR for the promotion of germplasm use through the development of a national data-base,
- ICAR institutes/state agricultural universities for the production of Breeders' and Foundation seed (including the crop research networks),
- central and state seed agencies (National Seed Corporation, NSC; and State Agro-Seed Corporations) for seed production, quality tests and labeling, and
- private seed agencies for the production of certified seed.

In 1988, 57% of arable land dedicated to crops with high-yielding varieties potential was planted with seed of modern varieties (CMIE, 1988). It seems, after a high increase of the amount of land under modern varieties at the beginning of the 1970s, that the amount of land under modern varieties is somehow stabilizing at around 55% (see Table 6.3).

In addition to the crops for which HYV are existent, there are other crops, lacking the research and development in new, higher yielding varieties. Consequently, only 30% of all arable land in India is cultivated with seeds of HYV (Mishra, 1996).

Apart from the known problems of innovation adoption, one reason for this was the restricted access of the private sector to the seed market. Until the late 1980s, less than 20% of the required seed was produced and marketed by the organized seed sector (ICR, 1995). Since October 1988, the new seed policy came into effect, liberalizing the import of seeds and planting materials. The import duty on seeds was reduced from over 90% to 15%. Additionally, the procedures for import and clearance have been simplified. Furthermore, measures were taken to encourage the production of more seed by the domestic seed market. The public sector still supplies about 70% of all distributed seed at present, whereas the private sector supplies the other 30% (Sharma, 1996).

Table 6.3. Trends in high-yielding varieties in Indian agriculture

Year	1970 /71	1980 /81	1982 /83	1983 /84	1984 /85	1985 /86	1986 /87	1987 /88
Area under HYV in million ha	15.5	43.1	47.5	53.7	54.2	55.5	54.1	55.7
% of HYV to total area under respective crops	17	45	49	54	56	58	56	57

Source: CMIE, 1988

Regardless of the obstacles for the private seed sector, India has a potential for seed production. Because of equally similar breeding technologies but lower wages, breeding may be more cost-efficient in India than in some industrialized countries. Apart from the multi-national breeding companies, which are cautiously entering the Indian market since the liberalization of the Indian seed policy, there are national private seed companies emerging, with a good deal of pioneering spirit.

Being the major starting point for an increased private involvement, the liberalization of the seed market seems to be the point of offence for some of the breeders involved in the public breeding sector who suspect a loss of influence and power to the private sector.

Although rather hesitant concerning a legal protection for crop varieties, India has been one of the leading countries in investing in crop improvement during the last 25 years (Jain, 1992). Neighboring countries like Bangladesh and Nepal are benefiting from the 30 to 35 varieties released annually in India, because these varieties are often cultivated there as well (Jain, 1992). Hence, a protection system would benefit India's investments and success.

Because of the inability in securing intellectual property, the private seed sector is mainly involved only in seed multiplication. In India, research and development for the site-specific problems are carried out only in hybrids, because of the inherent intellectual property protection (Kush, 1996).

Private breeding companies are interested in genetically coded information through the conserved accessions in the genebanks. Their interest lies in the reduction of their search costs, which necessitates a good information system on the supply side. However, at present the rice breeding companies conduct more cooperative activities with IRRI than with the national genebank in India, because of the lower search costs at IRRI (Narayanan, 1996).

Summarizing the Indian activities for conservation and utilization of PGRFA, India represents a centralized PGRFA ex situ conservation system. In addition to the formal system, different active groups involved in some conservation issues are constantly emerging. Two of the more important NGOs have already been introduced above. Both the MS Swaminathan Foundation and the Honeybee Network combine scientific research and practical work in the field of PGRFA conservation. Other groups and activities could be named as well.

The main in situ conservation activities in India, as well as in Germany, are coordinated and implemented by different agencies and ministries than other PGRFA conservation activities, which are coordinated and implemented by the Ministry of Agriculture. This is evidence for the fact that in situ conservation was emerging as the major conservation method for wild species of flora and fauna. Hence, in situ conservation came to fall under the responsibility of the Ministry for Environment. It will be of future importance either to merge all efforts for in situ conservation activities related to agriculture and integrate them under NBPGR or to keep the cooperation between the different ministries and offices going at a high level.

India is taking the lead position in formulating a sui generis system for the protection of property rights, in harmony with TRIPS and the Farmers' Rights of the International Undertaking. Two instruments will have to prove their viability: the introduction of the remuneration rights and the implementation of the National Community Gene Fund. A crucial point in the future will not be the payment into the fund, but rather the distribution of the financial resources collected. The example shown in the German case study indicates a possibility of collecting financial funds for some protection rights.

6.2 German Case Study[78]

Germany can be divided into three main ecological zones; 48% of Germany's total area (356,958 km^2) is classified as agricultural area (CRG, 1996). Parts of the remaining territory are protected with legally binding effect as protected areas. Additionally, since the UNESCO Man and Biosphere Program (MAB) came into being biosphere reserves were designated and wetlands of international importance were placed under the Ramsar Convention.

Germany's natural plant biological diversity is limited. Compared with tropical regions, the amount of different species is low. With the exception of some indigenous wild relatives from oats, rape, barley, potato, sugar and fodder beet, the important agricultural crops grown in Germany are not indigenous to the German flora. Only some vegetable and fruit species are indigenous to Germany. The main improving germplasm material for the fruit trees, however, originated in West Asia.

6.2.1 German System of PGRFA Conservation

In Germany, collection and ex situ conservation of PGRFA was started in the 1920s. The breeding research institutes took the leading role in the initial phase. From the start, the decentralized structure, which is typical for German research, was the main characteristic of all conservation efforts. Hence, in the early 1970s, several German institutes were active in the conservation of PGRFA, for instance: the MPI for Breeding Research, FAL, IRZ of the BAZ and the Teaching and Experimental Station for Integrated Plant Cultivation Güterfelde today's IPK (CRG, 1996)[79].

It was, however, not before 1986 that the Federal Ministry of Food, Agriculture and Forestry (BML) convened a project group to outline a concept for the conservation and use of plant genetic resources for the Federal Republic of Germany. The national discussion focussed on the conservation of PGRFA also because of the generally increasing awareness for the protection and conservation

[78] Important information was received through the discussions with the German Secretariat at IGR during the preparatory process and the German Country Report submitted to FAO.

[79] See the abbreviations list for the abbreviations in the text.

of nature. Furthermore, the international discussion and development concerning PGRFA conservation picked up speed, resulting in the establishment of the Commission on Plant Genetic Resources for Food and Agriculture as well as the adoption of the International Undertaking on Plant Genetic Resources (both in 1983). In 1990, the concept was presented to BML, but it has been implemented only partially since then (Bommer, 1995).

The reasons for the incomplete implementation of the concept are the decreasing availability of funds on federal and state level and the decreasing importance of the conservation of PGRFA and general conservation issues to the population. Furthermore, the concept was outlined for West Germany; through the German unification, the basis had changed and the sensitivity of structural changes had partially paralyzed the further implementation.

The establishment of the National Committee for the preparation process, the preparation and the accomplishment of the International Technical Conference on Plant Genetic Resources in Leipzig, as well as the continuously active role of the Information Center of Genetic Resources (IGR) has stimulated the implementation process again.

In situ conservation, understood as on-farm management, is still in ist infancy Germany. This is because few indigenous varieties exist. It is, however, also a political problem, because under the Plant Variety Protection Act, varieties have to be taken from the market after their variety protection and registration have expired. An in situ pilot project is currently being carried out for old varieties of wheat, rye, and potato, of which the seed and planting stock originates from the genebank in Gatersleben (Hammer, 1995).

Fig. 6.3. German system of PGRFA conservation

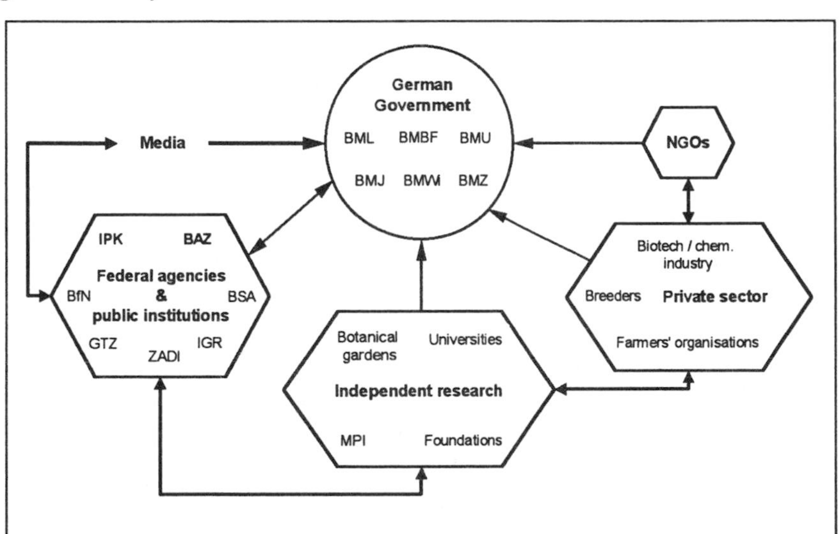

Because of the decentralized development mentioned above, the German structure of PGRFA conservation and management is characterized by many actors who influence the process on different levels. The actors and their influence on decisions relating to the conservation and management of PGRFA will be described in the following section and are depicted in Fig. 6.3.

Germany shows a decentralized structure through the legislative competence of the federal states for certain policy areas and partial tasks. Joint competence exist for nature conservation, agricultural structure and research policies of the Federal Government and the federal states. Even though the conservation of genetic resources falls under both legislation structures, the coordination for genetic resources conservation takes increasingly place at a national level. Hence, the *federal government*, plays an increasingly important role in all activities related to the conservation and utilization of genetic resources:

- The conservation and utilization of PGRFA is expected to contribute significantly to the future food security and to the agricultural improvement of the production conditions. Therefore, the Federal Ministry of Food, Agriculture and Forestry (BML) has the largest share in the conservation activities regarding the conservation and utilization of PGRFA. The Federal Ministry is leading the German delegation in international negotiations at FAO, mainly revising the International Undertaking on Plant Genetic Resources. BML is, however, able to achieve its objectives in PGRFA conservation and utilization on a national and an international level only in close association with other federal ministries.
- For instance, the implementation of any in situ conservation activities must be coordinated with the Federal Ministry of the Environment, Nature Conservation and Nuclear Safety (BMU) and its subordinate departments and institutes. BMU is taking the lead in the negotiations in most international fora concerning the environment - mainly CBD and its follow-ups. It is responsible for the implementation of CBD on the federal level. One important result of BMU's work in the field of implementing CBD is the German CHM Working Group on the Clearing House Mechanism (CHM) under the Convention on Biological Diversity. This group, consisting of representatives from BMU, GTZ, UBA, BfN, ZADI/IGR, University Bonn, German NGOs, and the private sector, and representing the National Focal Point to the CHM, has provided a concept for the development of the international CHM (BMU/CHM, 1997). BMU is generally aiming to integrate environmental protection into other action and policy areas - especially agricultural policies.
- BML depends on the expertise of the Federal Ministry of Justice (BMJ) especially for negotiations at international fora. Above all, the negotiations concerning the International Undertaking on Plant Genetic Resources and Farmers' Rights are accompanied by an expert of BMJ.
- Special attention must be drawn to the negotiations within the scope of WTO and TRIPS. The outcomes in these negotiations are influencing the conservation issues of BML as well as BMU. The Federal Ministry of Economics (BMWi), however, takes the lead in these negotiations.

- On the other hand, the Federal Ministry of Economic Cooperation and Development (BMZ) is responsible for the actual cooperation with developing countries. Consequently, close cooperation between BML and BMZ, and jointly with all other mentioned ministries is necessary for issues relevant to genetic resources conservation.
- Finally, the cooperation with the Federal Ministry of Education, Science, Research and Technology (BMBF) is of national importance. BMBF is improving the scientific basis of genetic resources conservation by financially supporting research related to conservation issues.
- Additionally, the federal states play a vital role in the conservation of genetic resources because of the German federalism. Besides their responsibility in main policy issues, they must implement the federal laws. For instance, BMU is responsible for launching the national strategy to implement the requirements of CBD. The actual power concerning conservation issues and their implementation, however, lies within the federal states.

Most of the activities relating to the conservation and utilization of PGRFA are carried out by public institutions connected to BML or to some other ministries. Some other activities, even those not directly correlated to the ministries, are sponsored by the government and are accountable to the government. In addition to the federal and states ministries, many *federal agencies and other public institutions* exist which are actively involved in the conservation process and influence the decision making process through their work and their output. All of these institutes are linked in different degrees to the federal or state ministries through financial or other support and cooperation:

- For instance, the Federal Agency for Nature Conservation (BfN), mainly involved in providing scientific advice and administrative support in conservation related issues to BMU, is responsible for the implementation of the CITES agreement. Furthermore, it administers federal funds for conservation projects in Germany (GCR, 1996). Another national agency working in the field of environment protection is the Federal Environment Agency (UBA). It supports BMU with scientific research, mainly in the field of quality and impact studies on water, soil and air.
- Furthermore, according to Germany's federal structure, in which universitarian research underlies the sovereignty of the individual federal state, the federal government maintains a series of federal centers for specific research. The Federal Center for Breeding Research on Cultivated Plants (BAZ), in the scope of BML, hosts one of the two main national genebanks, the Braunschweig Genetic Resources Center (BGRC) with approximately 53,273 accessions, of which 6,924 have been provided to users worldwide in 1996 (BAZ, 1997). This amounts to a respectful 13% of all conserved accessions, documenting a comparably high utilization rate (see Chap. 5.4). BGRC's main objectives are the safeguarding of agrobiodiversity and the supply of germplasm and information for research and development in the national and international context (BAZ, 1997). BGRC's special achievement is the use

of computer aided data documentation and research in information methodology (Frese, 1996). Furthermore, there is an important ex situ collection of vine at the Institute for Breeding of Grapes of the Federal Center for Breeding Research (BAZ) as well as some forest genebanks (GCR, 1996).

- The Information Center of Genetic Resources (IGR) was established in the course of organizing the German concept of PGRFA conservation in 1991. Its main tasks are information, consultation, coordination and research on genetic resources, cooperating with national and international organizations and institutes in the field of PGRFA (GCR, 1996). As the information center for genetic resources, IGR is one of the focal points in assembling ideas, research results, outcomes of workshops and international meetings and conferences and consequently making headway for further conceptualizing German conservation activities. IGR played an important role as the Secretariat for the National Committee for the Preparation of the ITCPGR as well as in the establishment of a databank for all conserved PGRFA (GENRES). IGR cooperates closely with the BMU through the BfN to support the development of the German CHM under the Convention on Biological Diversity (GCR, 1996).
- IGR's work must be seen in the context of the Center for Agricultural Documentation and Information (ZADI), another institute in the scope of BML, focusing on the management of national as well as international agricultural information resources (ZADI, 1995); IGR is one of its three departments.
- As an independent department accountable to BML, the Federal Office of Plant Varieties (BSA) has two main tasks: the approval of new varieties and the control of the seed market. The seed certification and market control are based on the national Seed Trade Act and the national Plant Variety Protection Act as well as on the international guidelines, laid down by the International Union for the Protection of new Varieties of Plants (UPOV) and agreed upon by its 30 member states. Besides UPOV, international cooperation exists in the EU, OECD, and FAO (BSA, 1996). Hence, although BSA is confronted with the output of breeding new varieties, BSA transfers varieties to the genebanks after the variety protection and registration have expired (after 25 years). Along with the old variety, the genebanks receive valuable evaluation data from the BSA (GCR, 1996).
- The other German genebank, the oldest and at the same time the largest storage facility for PGRFA in Germany with approximately 103,000 accessions stored at the Institute of Plant Genetics and Crop Plant Research (IPK), is supported jointly by the Federal Ministry of Education, Science, Research and Technology (BMBF) and the respective state governments[81], because it is classified as an institute of overriding importance[82](GCR, 1996). Its objec-

[81] There are some state governments involved, because besides the central genebank at Gatersleben, further collections are in some field offices in different states: Saxony-Anhalt, Mecklenburg-Western Pomerania and Saxony.

[82] These institutes are commonly named the "blue list institutes".

tive is the collection, conservation, multiplication, characterization, evaluation and documentation of material as well as its supply (Hammer, 1995). The genebank's crop evolution research has benefited from its close integration in IPK's work in the field of genetics, molecular and cell biology as well as taxonomy (Hammer, 1995). The Institute's comprehensive herbarium is outstanding in its uniqueness. Specialized in crops and their wild relatives, it is one of the largest herbaria with over 300,000 samples (Gäde, 1995). Besides the cooperation with other national and international institutes in the field of germplasm and information exchange, the genebank has started to support training programs for students coming from developing countries in cooperation with German foundations, e.g., DSE (Hammer, 1995).

- The German Technical Cooperation (GTZ) is as a public benefit service enterprise in the field of development cooperation, which supports partner countries development and reform processes. Some of the projects and programs are in the field of environmental protection and resource management. For instance, GTZ supports developing countries to fulfill their obligations under the terms of the Convention on Biological Diversity with the program "Implementing the Biodiversity Convention". Hence, being accountable to BMZ, GTZ's activities demand a close cooperation with BMU and its executive bodies as well as with BML and international active institutes like BAZ's genebank (BGRC) and IGR.
- Foundations, such as the German Foundation for International Development (DSE) - as independent acting public institutions - support the efforts of PGRFA conservation and utilization, mainly through advanced training measures for a skilled partner staff. In addition to foundations, other institutions, like the German Academic Exchange Service (DAAD), also implement training programs, supported financially by BMBF.

As part of the public sector as well, and mainly dependent on the federal or state ministries of research and education, are the institutions of *independent research*:

- Basic research on the different aspects of biodiversity is carried out at the universities and the Max Planck Institutes (MPI). Whereas universities are mainly subject to the state and their financial support, MPI are financed mainly by the federal government. The research related to agrobiodiversity is mainly carried out in the field of applied plant breeding and biotechnology. Because of their independent status and combined with the decentralized German research policy, a dissemination of information exists. As for the agricultural research related to developing countries, this gap was bridged by the Council for Tropical and Subtropical Agricultural Research (ATSAF); its successor is the newly established Advisory Committee for Development Oriented Agricultural Research (BEAF). The committee's main task is to serve as focal point for all relevant scientists, concerning the exchange and expertise of information.
- Most of the approximately 90 botanical gardens and herbaria in Germany are associated with university research departments. Special projects, however,

like the Contribution of the German Botanic Gardens to the Conservation of Biodiversity and Genetic Resources - Assessment and Development Concept are combined projects: financed by BMU, accompanied by BfN. The project is implemented by the Botanic Garden of the University of Bonn for the German Association of Botanic Gardens. Furthermore, this project is a co-operation with the Botanic Gardens Conservation International (BGCI). (GABG, 1998)

More involved in the applied research and development of new products and their utilization is the *private sector*, representing groups with partially contrary objectives: the farmers' organization on the one hand, representing the demand side for new seed, and the seed companies as seed suppliers on the other hand. The demand side is, however, not uniform. The majority of farmers utilize seed as one part of an input package with the advantage of being able to safe seed for the next growing season. The seed change rate in German agriculture has been estimated at approximately 50%, i.e., 50% of seed is bought as certified seed from the market and the other half is re-used seed.

In contrast to the majority of farmers, most of the farmers who support the ecological farm movement are interested in utilizing and exchanging old varieties, which are not tradable in accordance to the German Seed Trade Act. A further distinction must be made on the supply side. Until recently, the German seed supply was in the hands of small to medium-sized companies, working closely together with the public research institutes. In the past decade, the multi-national biotechnology companies, mainly petro-chemical companies which evolved into companies including biotech-breeding, have been entering the German seed market.

Although the private sector is non-governmental, the actors of the non-governmental/non-profit organizations (commonly "the" *NGOs*) must be differentiated. Although being heavily involved in general nature conservation and environmental protection, the German NGOs' interest in the field of PGRFA conservation played a minor role. Some NGOs promoted a change in the federal seed policy, so to keep old varieties in the market. These NGOs were mainly backed by farmers in ecological agriculture. After UNCED, and in the light of the International Technical Conference on Plant Genetic Resources, the Forum Environment and Development was formed however (GCR, 1996). This forum is a joint initiative to follow up UNCED, concentrated on the coordination of NGO positions and policy formulation, stimulating and providing studies and workshops on critical biodiversity issues and strengthening public awareness (GCR, 1996). Their main attention is focused on the international and national implementation process of CBD. The forum is financially supported by BMU through a project agency responsible for the coordination of the Forum (GCR, 1996).

Besides the interaction on the different levels in Germany, it is mainly the government through its respective ministries which is engaged in international collaborations. Above all it is mainly the participation in international fora, like the Global System of FAO, CBD and GEF as a funding mechanism for environmental programs and projects, as well as initiatives of UNESCO (Man and Biosphere Program), other conventions such as the Ramsar Convention on the conservation of wetlands of international importance, and the CITES Convention cooperation in PGRFA conservation is the intensive collaboration with international agricultural research, the research centers of the Consultative Group on International Agricultural Research (CGIAR) as well as non-CGIAR centers.

Because of the federal structure of German policies, the decision-making process concerning the German conservation strategy incorporates the interests of various stakeholders at the different political levels. This is achieved by circulating a final strategy paper to the members of Parliament, the federal ministries, the federal governments, NGOs, and the interested public via media. Additionally, the federal governments disseminate the paper to district- and county-level officials. (Auer, 1995). This is a time-consuming, but participatory approach to include all relevant interests in the decision-making process. The approach is linked with the expectation of easing the implementation process.

The establishment of working groups is an additional instrument to involve the different institutes with their different point of views and expertise in the decision making process on a national level. For instance, the Working Group for the "Conservation of Forest Genetic Resources" was established, integrating the different actors concerned on a national and state level (GCR, 1996). The concept for the conservation of biodiversity is another example, showing how different expertise from different areas and levels come together and generate a concept. The general concept is based on the Federal Nature Conservation Act, designed by BMU. The scientific background for the concept of nature and species protection, however, is derived mainly from the Federal Agency for Nature Conservation (BfN) as well as the different states and their scientific institutions.

A closer evaluation of the efficiency of the German conservation program together with programs of other countries is undertaken in Chapter 5.4. Hence, here only a close look will be taken at the genebanks' integration into the system of utilization. The two national genebanks are different as regards their historical backgrounds, their present conservation strategies and their funding. In analyzing their contribution to the utilization of PGRFA, they are, however, similar. The supply of conserved PGRFA (in terms of accessions), reflects the degree of accessibility of the storage facility on the one hand and it indicates the value of a whole collection on the other hand. While IPK has provided an annual average of 12,252 samples between 1953 and 1992 (Hammer et al., 1994), BAZ has provided some 7,000 samples or more annually[82] (BAZ, 1997). Compared to their overall conserved accessions, IPK supplied 12% and BAZ 13%, which is a good result for their accessibility in comparison to other genebanks (see Chap. 5.4).

Two further points must be highlighted. First, the supply of accessions to meet domestic and foreign demand is well-balanced in both genebanks, whereby IPK, with 65%, provides slightly more accessions for the "domestic market" than BAZ with 56% (GCR, 1996). In spite of being national genebanks, this indicates that international cooperation is intensively utilized and their international reputation is good enough to attract a significant demand. Furthermore, it is interesting to note that although Germany is an agrobiodiverse-poor country, the two main

[82] Since 1982, the annual supply has been constantly over 6,000, with peaks in the year 1986 (over 14,000), 1992 and 1995 with over 10,000 samples (BAZ, 1997).

conservation facilities provide more accessions to the "international market" than they receive from foreign sources. IPK, for instance, has registered an annual average of only 2,431 entries into the genebank since 1953 (Hammer et al., 1994). According to Gäde (1995), 41% of the accessions received from outside the genebank in the years 1991 to 1993, were from outside Germany. However, this does not imply that the 59% of all new entries into the genebank were of German origin. They are made up in part by newly collected material, coming, however, predominantly from another German institute.

The expenditures for all activities related to the conservation management of PGRFA in Germany are approaching US $ 113,215 million per year[83]. These expenditures involve very different costs, which have to be discussed in the following section (see Table 6.4).

In analyzing Germany's expenditures for the conservation management of PGRFA in 1995, one major point can be highlighted immediately - although it was not possible to assess all expenditures on a federal and state level, as well as those expenditures of all organizations involved in the conservation management. The expenditures for pure conservation efforts constitute only a fraction of the overall expenditures. Only approximately US $ 4 million are listed in Table 6.4 for the conservation of germplasm in genebanks and other storage facilities. The two individual genebanks, IPK and BGRC, are, however, responsible for only 60% of the stored accessions in Germany; other material is stored in the working collections of various institutes. It has been moderately estimated that approximately US $ 11 million are spent for conservation in the narrow sense of solely conserving germplasm[84] (Oetmann, 1996). Approximately 10% of all German domestic expenditures for PGRFA conservation management which are spent for the long-term storage of PGRFA in the narrow sense reflect the imbalanced expenditure structure between pure storage and processing costs for PGRFA. With approximately 500,000 accessions, the USA spends US $ 20,000 annually for the national conservation system, which does not imply major conservation management activities. Hence, based on the average expenditures per conserved accession, Germany and the USA spend approximately the same amount on pure conservation activities. The further involvement in information processing is, however, of very high importance for Germany and other European countries. Hence, 45% of the expenditures is dedicated to improving the accessibility and utilization of stored accessions by evaluation and prebreeding activities. As has been shown at various points throughout this study, the opportunity for utilization decreases significantly by simply storing PGRFA in storage facilities. Hence, the accessions stored in genebanks and other facilities are worth demanding only by including some information management, i.e.,

---

[83] In the preparation process for the 4th International Technical Conference on Plant Genetic Resources Oetman (1996) gathered the German expenditures from the various sources; for this study they were then compiled differently.

[84] The quoted US $ 11 million are not clearly marked in Table 6.4, they are mainly hidden in the expenditures for breeding, but because of joint costing in these institutes, the plain conservation expenditures could not be separated.

characterization and evaluation data, and some initial information processing, i.e., prebreeding (Hammer, 1995). Therefore, the expenditures for documentation and evaluation as well as elements of the public breeding activities were integrated in the calculations.

In addition to expenditures for conservation and partly for breeding, the institutional expenditures were integrated into the calculations. These mainly consist of expenditures for IGR and the Federal Office of Plant Varieties (BSA).

Table 6.4. German domestic expenditures for the conservation management of PGRFA in 1995

Origin of expenditures	Itemized expenditures in [US $ '000]	Total expenditures [US $ '000]	Portion of total expenditures in [%]
Genebanks:		4,140	4
IPK (Institute of Plant Genetics and Crop Plant Research)	3,450		
BGRC (Braunschweig Genetic Resources Center)	690		
Conservation management in plant breeding:		51,164	45
BAZ (Federal Center for Breeding Research on Cultivated Plants)	30,360		
MPI (Max Planck Institutes)	13,800		
Research in Genetic Engineering	2,864		
Federal Research Center for Forestry	4,140		
Institutional expenditures for conservation management:		34,673	31
IGR (Information Center of Genetic Resources)	173		
BSA (Federal Office of Plant Varieties)	34,500		
Financial support for PGRFA related Projects		23,238	21
Domestic expenditures total:		113,215	

Source: compiled according to data from Oetmann, 1996

As can be seen in Table 6.4, IGR's share of the expenditures in 1995 were less than 1%, whereas BSA contributed approximately 30% to the overall expenditures because of its manifold activities. Not all of BSA's work contributes directly to the conservation management of PGRFA in Germany. Hence, the expenditures of BSA for PGRFA conservation management are overestimated to a certain extent.

Furthermore, expenditures consisting of financial support for PGRFA related projects were also integrated into the calculation. These supports for PGRFA related activities were financed by nearly all of the ministries mentioned above: BMU, BMWi, BMBF, and BMZ.

The total foreign assistance contributed by Germany for the conservation management of PGRFA amounts to approximately US $ 18 million. These contributions include, however, expenditures directly correlated to PGRFA conservation management, such as the support for the Kenyan genebank, but also partially Germany's contribution to the CGIAR system, which might be utilized for PGRFA conservation management. Hence, this figure underlies a significant variation.

Summarizing the analysis of German expenditures for PGRFA conservation management, it may be highlighted that the conservation of PGRFA in a narrow sense is only a minor item in the overall expenditures amounting to only approximately US $ 11 million. This means a unit cost of US $ 55 per accession conserved in German genebanks - which is still higher than the annual unit costs of US $ 37 in the USA[86].

6.2.2 Breeding

The breeding and marketing of seed in Germany depends on two federal laws: the Plant Variety Protection Act, which determines the intellectual property of new breeding products, and the Seed Trade Act, which sets the threshold for the market entrance of a new variety as well as the definite end of the variety.

The intellectual property rights for breeding are regulated by the Plant Variety Protection Act of 1985. Every new, homogenous and stable variety, which is distinct from other varieties, may be granted variety protection by the Federal Office of Plant Varieties. The owner has the exclusive right to breed, multiply and sell the variety for as long as the variety protection is valid (generally 25 to 30 years). After the expiration of the variety protection, the variety is free for use by everyone. A prolongation of variety registration according to the Seed Trade Act is, however, necessary for further marketing, because of an appropriate cultivation and market importance of the specific variety. After the expiration of the variety registration, the variety must be removed from the market. (BSA, 1996).

---

[85] The quoted US $ 11 million are not clearly marked in Table 6.4, they are mainly hidden in the expenditures for breeding, but because of joint costing in these institutes, the plain conservation expenditures could not be separated.

[86] See for more detail Chap. 5.2.

This procedure is the main reason for the small share of old varieties in German farmers' fields.

The Plant Variety Protection Act, which determines the exclusive rights for the sale of a variety, grants two exceptions: breeder's exemption and farmer's privilege. Both exemptions have their supporters as well as their opponents. The biotechnology industry invests heavily in research and development and is very much interested in protecting their inventions against competitors. The breeding companies, however, depend on the free access to prebred products for further breeding (Hammer, 1995). Without utilizing the new products from other companies or from the public sector[87], the breeding companies are too small to invest much financial resources in basic breeding research (Hammer, 1995). The typical German small and medium-sized breeding companies as well as the big biotechnology companies refuse the farmer's privilege, claiming that it is an obstacle and disincentive for further breeding efforts. The farmers and many NGOs, however, understand their privilege as a kind of traditional fundamental right (BUKO, 1996). The re-used seed, derived from farm-saved seeds is still used in significant quantities. It is estimated that approximately 50% of the area under cereals (excluding maize) is planted with re-used, farm-saved seed in Germany; for potato and grain legumes, the seed change ratio is 40% and between 30 and 50% respectively (BML, 1994).

Germany has tightened the variety protection since 1997 by introducing a fee for the re-use of farm-saved seeds, whereby smallholders may still benefit from the farmer's privilege. This modification is based on the changes in UPOV's convention from 1991 (UPOV '91). The underlying principle of the agreement between the German Farmers Association and the German Plant Breeders Association is an increase in the farmers' seed quality by the promotion of certified seed and the reduction of the economic advantage and profitability of re-used seed through a fee for the re-use of seed.

The amount of the fee depends on the farmer's seed change ratio (SCR)[88]. Farmers with a SCR of over 60% for cereals and legumes and a SCR of over 80% for potatoes, are released from the fee. Lower ratios lead to fees between US $ 2 and US $ 13 per hectare for cereals/legumes and between US $ 56 and US $ 83 per hectare for potatoes. The fee is levied as a lump sum and its amount depends on the SCR and is fixed every year. The lower the SCR, the higher the fee. In contrast, farmers obtain a discount of 10% on the license fee for a specific certified seed if their SCR for that variety is over 80%[89]. (DBV/BDP, 1996).

The agreement signed in 1996 must be viewed in light of the above mentioned seed change ratio of approximately 50% and its slight upward trend in the last years (BML, 1994). The promotion of the breeding efforts and weakening of the re-use of seed are attempts made to increase the use of technological development on the farms. The major problems with this new system, however, are the

---

[87] For instance, the universitary breeding research.

[88] The seed change ratio (SCR) is defined as the ratio between the share of area planted with certified seed to the overall area planted to the respective crop.

[89] The discount would amount to approximately US $ 0.6 per 100 kg of certified seed.

definition of a smallholder ("Kleinerzeuger") and the institutional process of imposing the fee and the connected transaction costs.

A smallholder is defined according to BDP (1997a) as a farmer cultivating less arable land than is needed for the production of 92,000 kg of cereal. By an average of 575 kg of cereal per hectares this is 16 hectares[90]. By implementing this criteria, between 250,000 and 350,000 of approximately 553,000 farms in Germany[91] will be exempted from any fees for the seed re-use[92].

The fee for the re-use of seed will be collected based on a system of self esteem and the provision of evidence. This procedure will be the crucial point of the implementation. With an average annual fee from approximately US $ 21 up to US $ 136 per farm, the collection costs may reduce the net benefit by 50% because of administrational and enforcement costs. On a larger farm the fee can, however, easily amount to US $ 520 reducing the costs significantly[93]. The only sanctions which may be imposed in case a farmer is not cooperating is to charge a fee - four times the amount of the license fee of certified seed[94]. Resources have to be made available to monitor the agreement; 627 different varieties of the different relevant crops are protected by the Plant Variety Protection Act (BSA, 1996).

Although in the worst scenario a zero play because of the high enforcement costs, the significance of one incorporated benefit must be emphasized: the fee for the seed re-use is not only an additional income for the breeding industry, but it also reduces the comparative advantage of the seed re-use. This implies an increasing number of farmers who will increase their SCR. By an average SCR of 50% for cereals on German farms, it may be predicted that the majority of farmers will increase their SCR by approximately 10% in order to avoid any seed re-use fees at all.

Because the agreement was first implemented in 1997/98 no trends can be identified yet.

One major determinant of the decline of agrobiodiversity in Germany is the Seed Trade Act. It restricts the market entrance only for those varieties, which meet the criteria of distinctness, uniformity, and stability. Furthermore, it restricts the market availability of a variety to a certain period of time. Both restrictions were put into force to protect the consumer, i.e., the farmer. At present, the call for a change in the Seed Trade Act by liberalizing the legal structure is increasing, e.g., by the Forum Environment and Development (BUKO, 1996). Hence, BSA is

---

[90] For potatoes it is defined to be less than 185,000 kg, i.e., around 5 hectares.

[91] Only farms with more than 1 hectare were included.

[92] Approximately 250,000 farms have less than 10 hectares, an additional 100,000 farms have between 10 and 20 hectares. (BDP, n.d.).

[93] If we assume an average farm size of 30 hectares, of which 70% is planted with cereals, estimating the average SCR in German agriculture of 50%, the farmer will be charged a fee for seed re-use for 10.5 hectares. The total of the fee, depending on the kind of cereal used, will therefore amount to US $ 21 up to 136. However, on a farm with 130 hectares, the fee can easily amount to US $ 520 (BDP, 1997c).

[94] It is US $ 6 for 100 kg of cereal, assuming a need of less than 200 kg seed for one hectares, the licence fee is less than US $ 22 per hectare. Hence the punishment is more than ten times the amount of the fee for seed re-use.

ready to agree to an additional way of market entrance for specific varieties, either landraces or old cultivars (AGRAR-Europe, 1997). The fundamental principle for market entrance is based on the minimum standards of germination, health, and purity and should be labeled differently than the normal certified seed. If the Seed Trade Act will be modified according to this line of reasoning, it would enable the consumers (i.e., the farmers) to choose between a larger number of varieties with different quality standards. This might increase the diversity of varieties, especially in the fields of farmers in the ecological farm movement.

6.3 Summary of the Indian and German Case Studies

To compare the German national system with the Indian system, the main difference is the decentralized German concept. This is mainly the result of the overall federal structure. One common institutional problem which occurs in both countries is the separation of the in situ conservation programs from all the other public conservation activities.

Like India, which is taking the lead in the formulation of a sui generis system for property rights, Germany is taking a step forward in partly abolishing the farmer's privilege. The implementation of both actions will have positive effects on the research and development activities of the private breeding industry. In India the implementation may prepare the ground for the final break through of private breeding, while harvesting the benefits of public breeding investments of decades. In Germany the implementation of the new agreement may safeguard the medium-sized breeding companies against the increasing competition from the biotechnology companies. As India's sui generis system will include the concept of Farmers' Rights and will serve as an example for other countries, it may be of interest to follow the implementation of the newly agreed upon fee in Germany. Especially the new way of collecting the fee will have to prove its economic feasibility.

Finally, while in Germany the rights of farmers are reduced in the interest of the medium-scaled breeding industry and for the benefit of an advanced technology transfer between breeders and farmers as well, the Indian government is attempting to implement a property right in the interest of the private sector and as well as being for the benefit of marginalized farmers.

7 Conclusions for PGRFA Conservation Policy and Further Research Issues

Although agrobiodiversity is a subsystem of biodiversity, differences exist between the two, for instance, in the underlying causes for extinction as well as in the transaction chains between collection and ultimate producer and in the revenue-generating potential of products. Hence, economic issues of the conservation of agrobiodiversity must be treated separately from those of biodiversity in general.

7.1 The Conservation and Service Center as a Focal Point for PGRFA Conservation Management

It has been often stated that PGRFA in farmers' fields is *"... being lost at a rapidly increasing rate."* (Esquinas-Alcázar, 1996, p. 3). These statements are based on studies identifying the loss of specific varieties at particular sites; for instance, the extinction of vegetable varieties of asparagus, beets, onion and others in the USA between 1903 and 1983 with extinction rates between 87% and 98% (Fowler and Mooney, 1990). This study illustrates, however, the fundamental problem of defining agrobiodiversity. In principle, agrobiodiversity can be defined as the diversity on a genetic level, variety level, and species level. Because of lack of technology needed for the measurement of genetic diversity, most studies differentiate agrobiodiversity on an intra-species level. Taking the amount of varieties as the basis for the decline of agrobiodiversity, there is a change in agrobiodiversity. For the most part, the area under a diverse set of crops and varieties is decreasing. The actual loss of varieties such as those mentioned above, cannot, however, be proven in many places because of lack of information. If information is available, it is often inconsistent[95]. Detailed studies for identifying genetic diversity at species level do not always prove the thesis of a rapid decrease in agrobiodiversity (see Smale, 1997, Tarp, 1995). The situation of PGRFA in

[95] For example, FAO (1993a) reports that 75% of India's agricultural area which once accommodated up to 30,0000 rice varieties is now taken up by only 10 varieties, but that on the remaining 25% still all varieties exist. The contrary is reported in the Indian Country Report to FAO, according to which only 16,000 rice varieties are still cultivated by farmers in India (ICR, 1996).

farmers' fields is characterized by a decline, but one which may be less rapid than expected; many varieties may reach the threshold of extinction soon while others may have reached this threshold already.

The marginal costs of losing one more crop variety, however, have not yet been determined, depending essentially on the variety specific threshold. The main value of PGRFA - the use value as (raw) material for breeding activities - can be identified or estimated by its close correlation to the distribution of germplasm to breeders. Although some countries have a distribution rate of over 10%, the majority of genebanks do not provide very much of their stored PGRFA to breeders[96]. The reasons for ignoring the stored genetic resources are the low quality of the raw material and – increasingly so - the limited willingness to accept legally binding agreements.

In the light of signs that crop variety is declining, and despite the uncertainty in the decline of genetic diversity, and in view of vague but raising estimations concerning the value of PGRFA, there exists an overwhelming international political intention to conserve PGRFA. There are two implications in the process of realizing this purpose:

Further research is necessary to assess the value of PGRFA. The benefits of conservation must be analyzed more in depths to guide the conservation investment on national and international level. As discussed in Chapter 3.2. the related work done differs widely in its results. The values attributed to PGRFA have often not been separated from the incorporated value through the breeding activities. It is without question that a high value is attached to the breeding activities which include the raw material of PGRFA. The separation of the joint value is the precondition for a rational debate over the internalization of the external benefits ("benefit sharing") of PGRFA.

It is necessary to define the conservation objectives for adjusting these to some cost-efficiency criteria. As long as the benefits of PGRFA conservation cannot be better quantified, the conservation investments must follow the criteria of cost efficiency. The definition of objectives is, however, necessary for assessing the efficiency of conservation activities. Furthermore, the objectives must be quantified as well to be able to carry out the assessment as transparently as possible[97]. Besides the conservation of genetic resources for future utilization and their adaptation to changing conditions, one important conservation objective is the accessibility of germplasm for present needs in research and development. The close link between conservation and utilization can be assessed by the distribution rate, characterizing the accessibility of germplasm. The present distribution rates of most genebanks are not convincing. The low distribution rates reflect the low

[96] Distribution rate defined as the rate of distributed accessions to the stored accessions in a genebank (see Chap. 5.4). This ratio includes the distribution of accessions to other genebanks as well, i.e., the amount of accessions supplied to the (private and public) breeding industry is even less – it may be estimated to be less than 40%.

[97] See Chap. 5.4 for more detail.

breeding quality of the germplasm[98] on the one hand, and the poor quality of conservation management on the other. Hence, in order to supply the breeding market with high quality material, *the conservation institutes must process germplasm from raw material into high quality genetically coded information.*

In times of increasing privatization of the breeding activities, and therefore higher competition between breeders, and a continuously increasing demand for food produced on limited arable land, the breeding industry is looking for means to reduce the breeding cycle. Genetic engineering is one way of reducing the time span between the start of a breeding program and the final variety by incorporating only specific genetically coded information into the breeding line. The conventional breeding, however, lacking the capacity for genetic engineering, is able to reduce the length of a breeding program by making more selective crossings. Hence, both breeding techniques are in search of well processed genetically coded information. Presently, these are available in restricted numbers of the elite lines of breeders. If the PGRFA conserved worldwide could be processed, the supply of a huge diversity of reidentifiable and technically properly treated genetically coded information could be made available to the breeders and considerable willingness to pay for such information may be expected.

Following this argumentation, it is obvious that the storage facilities will have to play a more active role in the future. As the interface between conservation and utilization, the genebanks and their related facilities (e.g., tissue culture, cryopreservation facilities, and botanical garden) must bridge the gap between conservation and utilization by:

- improving their quality standards for the ex situ conserved material (health, regeneration etc.),
- improving the processing of the conserved germplasm (characterization, evaluation and prebreeding programs) as to transform PGRFA from raw material into attractive genetically coded information,
- coordinating all in situ conservation programs (from on-farm management to habitat conservation),
- surveying the area under landraces, especially the transformation of production systems in marginal areas,
- centralizing all national relevant information, and
- presenting all national storage facilities as one conservation and service center being the supplier of genetically coded information for the breeding industry.

It is of no importance whether the conservation center is made up of only one storage facility in a very centralized system (as in India) or as various storage facilities including research units in a more decentralized system (as in Germany); important is that the different conservation and processing efforts are coordinated and that the information is presented as a unit for the demand side (see Fig. 7.1.). It has to be the focal point for all information on ex situ and in situ conservation activities in the country as well as for any collaboration on international level.

[98] By integrating germplasm from a landrace in a breeding program, the whole breeding cycle is prolonged for about 10 or more years (Smith and Salhuana, 1996).

Fig. 7.1. A proposal for a PGRFA conservation and service center

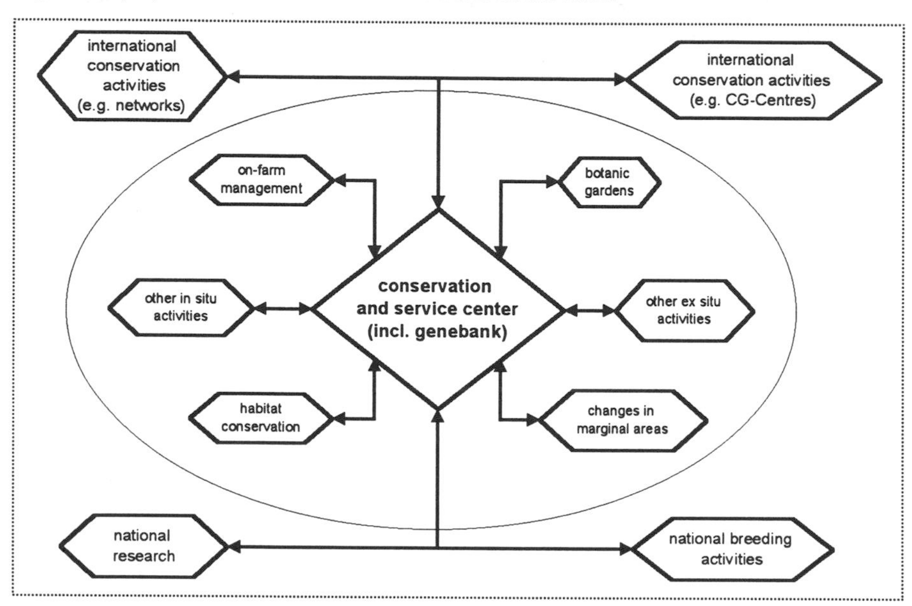

Such a system would help to reduce the search costs for the breeding industry to a minimum. In addition to the service aspect and the additional value added for the demand side, the processing task is advantageous for the protection of germplasm. The enforcement of any rights, as royalties or other internalization mechanisms is impossible without gathering all existing genetically coded information of an accession. The conservation and service center (CSC) can only claim any internalization of social benefits or any compensations if it can prove which genetically coded information was sent to which breeder. The CSC can act as a broker at a national level and will have contact to the clearing house on an international level.

7.2 Incentives of Moderation

No remuneration is offered at present; either for farmers who maintain agrobiodiversity in their fields, or for the conservators who are involved in the first processing of PGRFA. The storage facilities are trying to generate a data base containing information on their stored material in order to improve the accessibility of specific germplasm. They distribute the material and the respective information for free, but often do not receive new or additional information on the material distributed in return (Hammer, 1995). Many genebanks in developing and industrialized countries are suffering from insufficient budgets and are threatened by further budget cuts (e.g., Shands, 1994; Hammer, 1995; Chandel, 1996).

The farmers will go on maintaining their production system as long as their private marginal benefit is higher than their private marginal costs. They will maintain a specific level of PGRFA diversity with their traditional production system. By changing the production system to a "modern system", the diversity level will decrease. This transformation is determined by other criteria than the conservation of PGRFA. Hence, conservation of PGRFA is conducted at the farm level only as side-effect of other decisions; as a positive external effect. That is to say, no costs are associated with the conservation of PGRFA in farmers fields. The de facto in situ conservation reflects the economic principle of efficient resource allocation. Conservation takes place where diversity is high and opportunity costs are low. In situ conservation per se has opportunity costs at the national level however, which reflect the competition between utilized arable land for a high level of agrobiodiversity and for the increasing need to secure food demands. Consequently, in situ conservation of PGRFA should be promoted in a rather restricted way. Incentives for the conservation of PGRFA should only be applied, if the utilized arable land for a variety is falling below its specific areal minimum safety standard.

The "controlled in situ conservation" (Chap. 3.4.2.2.) fits a potential market solution in a clearly defined framework. This framework is determined by social benefits which cannot be internalized into a complete market system, i.e., the framework protects the non-market values of PGRFA against a market system. *A further detailed concept of this restricted in situ conservation incentive may be developed* based on the outline introduced in this study.

The comparative advantage of the landraces is shown wherever a higher level of agrobiodiversity exists without any additional incentives. It indicates that private benefits of utilizing a specific production system with a high level of agrobiodiversity are higher than with other production systems. It is important, however, that a free and unrestricted competition between landraces and new varieties exists, concerning the availability of seed, unbiased prices for inputs etc.

Concentrating on the national level, it can be stated that the social benefit of conservation as assessed by the international community may be higher than that of a specific country. This is the case in a country with a high level of agrobiodiversity but with a low level of technology for processing genetic resources. In other words, the standard of the country's public and private crop improvement programs determine the social benefit of a country's access to genetic resources. To promote ex situ conservation, incentives must be supplied to a country, only if its social benefits are lower than those of the international community.

Incentives are necessary for the continuation of PGRFA conservation, ex situ as well as in situ. It has been suggested, however, to give incentives only where necessary. The requirement of incentives depends on the adherence of a crop-specific areal minimum safety standard (for the in situ conservation) and of the difference between national and international social benefit considerations (for the ex situ conservation). Hence, *further research is necessary to outline operational ways of identifying the specific areal minimum safety standard for the different crop varieties as well as to estimate the social benefits at the different levels.*

7.3 Compensation or Internalization?

The present structure of the conservation and exchange system is changing, because of the emerging awareness of the value of PGRFA and the changes in the intellectual property rights. Countries with a high agrobiodiversity are calling for a share of the benefits derived from the utilization of PGRFA. Two aspects must be highlighted in this context:

"Benefit sharing" can be understood as an incentive for present and future conservation efforts at the national level (for the ex situ conservation) as well as at the farmer level (for the in situ conservation). *Incentives for the conservation of PGRFA are needed* because of the constant decline in PGRFA in storage facilities as well as in farmers' fields (Chap. 7.2.).

"Benefit sharing" can furthermore be understood as an equity issue in light of the positive external effect produced by farmers and its utilization by private (and public) breeders. *The suppliers of PGRFA are attempting to use their property rights and to derive benefits resulting from the supply of germplasm.* Countries do so by establishing a market power and threatening to close the access to their national germplasm pools. This attempt is supported by the Convention on Biological Diversity, which reaffirms the countries' sovereignty over their genetic resources. The agrobiodiversity-rich countries encounter two major obstacles in trying to enforce the exclusiveness of their PGRFA: first, most of the technology-rich and agrobiodiversity-poor countries hold important ex situ collections. Hence, these countries can feel"self-sufficient" for a long time in regard to the supply of PGRFA as raw material for their breeding programs. No agreement was reached on the legal status of ex situ conserved germplasm collected before the enforcement of CBD in 1993. Hence, all the accessions collected before 1993 (which constitute most of the conserved accessions), will be kept open to all users. The second obstacle for the emerging market power of agrobiodiversity-rich countries is the prohibitive transaction costs of controlling the export of germplasm or of the legal action later on[100]. Additionally, some agrobiodiversity-rich countries still depend on public or private breeding activities in other countries (see Chap. 7.4). Hence, at present, countries - especially those with high agrobiodiversity and low technological potential - are interested in supplying germplasm to the breeding industries in order to receive new varieties.

All governments are challenged now to modify existing property rights arrangements and build new arrangements for genetic resources so that the benefits can also serve as an incentive for those conserving PGRFA on the one hand, and the costs of the transactions between them and the users are not prohibitive on the other. As long as property rights are not at all defined, poorly

[100] It may be feasible at selective places, e.g., at the airport; but a control of all borders is impossible, especially bearing in mind that germplasm crossing the border can always be declared as food product.

defined, or badly enforced, adequate incentives for the conservation of PGRFA are not provided.

There are two alternative ways of implementing the idea of benefit sharing: either a system of compensation for the utilization of PGRFA or the internalization of existing or expected benefits. Different legal systems support different approaches. TRIPS represents the possibility to internalize the social benefits on the one hand, by developing some kind of protection system for genetic resources (sui generis). On the other hand, the concept of Farmers' Rights is aiming at a compensation solution testifying the moral right of farmers to be acknowledged for the past, present, and future breeding efforts.

In principle, the internalization would be the most efficient way of organizing and financing the conservation of PGRFA, if the transaction costs for developing and maintaining a conservation and exchange system are smaller than the resulting benefits. An appropriate response to the loss of biodiversity is the reduction of the size of the feedback loops so that those who make decisions about resources use, quickly feel the impact of their decisions (McNeely, 1996). This means, if the expected value of PGRFA is internalized, the decisions at the farm level and at the national level would take into account agrobiodiversity as a value. Hence, the fixed link between the level of PGRFA diversity and state of production system could be broken up. To reduce the size of the feedback loops in agriculture is, however, more complicated than for biodiversity in general. The feedback loop for the benefits resulting from one specific landrace will be too long for the normal time frame of an individual decision-making process on the farm level [100].

Consequently, the dilemma is obvious: the value of genetically coded information can seldom be determined a priori but only observed a posteriori, i.e., as a result of their success on the market. Hence, the internalization of benefits as royalties (a posteriori) for individuals is not possible because of the intergenerative structure of benefits. The internalization of benefits, as payment on account (a priori), will seldom reflect the true use value of specific genetically coded information.

A full internalization of the social benefits for the in situ conservation carried out by individual farmers is impossible. Hence, without an internalization, the incentive for farmers will not be high enough to conserve the social optimum. In order to maintain a minimum area under each crop variety in situ, the conservation and exchange system must provide mechanisms adequate to the situation sooner or later.

At a national level, an internalization can be realized in a bilateral exchange system. A complete internalization of the social benefits is, however, not feasible at the moment, because of the intergenerational aspect of benefit arising as well as the high transaction costs of enforcing a control system. Hence, a multilateral exchange system (MUSE) with some bilateral elements, based on the concept of compensation, seems to be the most operational system for the time being.

---

[100] The benefits may arise after 15 and more years, representing the normal time lag in a breeding program (Smith and Salhuana, 1996). In other words, the expected private benefit in 15 or more years will be discounted resulting in a very small or no present benefit.

Similar to the solution at national level, compensation will be the underlying concept for any implementation of Farmers' Rights. This concept can be understood as:

- the general compensation for farmers' past contribution by promoting modern varieties for today's farmers (agricultural and general economic development),
- the targeted compensation for specific farmers' present contribution to maintain specific (endemic) traditional varieties (e.g., controlled in situ conservation), and
- the targeted incentives for specific farmers' future contributions to improve traditional varieties (adapting landraces to changing conditions).

Hence, a compensation solution will be the most feasible. The crucial point of a compensation solution, however, is the question whether and how the compensation reaches the farmers. Furthermore, inevitable in the concept of compensation, there will be the attempt of agrobiodiversity-poor and technology poor countries to act as free riders.

7.4 Increasing Effectiveness Through a Market System, Technology Development and Cooperation

Conservation costs may be reduced by adapting conservation techniques to the different conservation objectives[101]. In situ conservation is undertaken by farmers as a positive external effect as well as through high cost programs. *Further research is needed to analyze in situ conservation instruments, their costs and above all their impact on the level of agrobiodiversity in the region.* Are these programs conceptualized to maintain a specific level of agrobiodiversity or to increase it? How would agrobiodiversity develop without a program?

Reflecting the high costs for in situ conservation programs, and taking into account the opportunity costs at a national level for the area under landraces, it follows that the area under landraces should be the result of a market competition between landraces and new varieties. For this it is necessary to abolish distorted incentives, like subsidies for the new varieties as well as for fertilizers and other inputs for the new varieties, while internalizing the degradation costs of natural resources (soil and water) [102]. The only necessary limitation in the system is the condition that all varieties must be kept cultivated. If the area under one landrace falls below the areal minimum safety standard, the CSC must intervene by implementing some controlled in situ conservation programs. Furthermore, to equalize the conditions for a market competition, it is suggested to support small-

[101] For instance, the low-cost cryopreservation may be utilized for the long-term conservation without any accessibility for medium- or short-term access. The access, in turn, may be achieved by low-cost short- and medium-term conservation methods.

[102] The internalization of degradation costs applies to new varieties as well as for landraces. It is not per se garantueed that landraces degrade less the natural resources than HYV do.

scale seed production and distribution enterprises where necessary in order to improve the access to a wider range of planting material, including the provision of landraces stored ex situ, and the promotion and improvement of farmers' breeding of varieties. In this connection it is of importance to review and adapt agricultural development policies and regulatory frameworks for variety release and seed certification. It must be, however, carefully considered how far the support for landraces leads to a perverse incentive for the new varieties which is unacceptable because of the overall need for food security.

The free competition between modern varieties and landraces limited only by the restricted areal Minimum Safety Standard (aMSS) will increase the effectiveness of in situ conservation, by reducing the area planted by landraces as well as guaranteeing the maintenance of all landraces. Furthermore, a free competition will identify the economic marginalized areas, and the crops and varieties of which the landraces still have comparative advantages. Hence, the attention to further development of infrastructure and for further breeding must be turned to these regions and to these crops and varieties in particular.

Agrobiodiversity-rich, but (breeding) technology-poor countries are in a dilemma[103]: the greatest possible value can be created within the country itself only by processing the collected germplasm to value-enhancing genetically coded information. The lack of technological capacity as well as human resources, however, prevent this and hinder the supply of high quality PGRFA on the world market. Consequently, these countries must seize their opportunity, wherever possible, to develop and improve their technological potential of breeding and support of this is also in an international interest.

The analysis of the *effectiveness of storage* facilities in 34 countries suggests that existing financial resources should be used more effectively. The efficiency of ex situ conservation may be improved through the rationalization of efforts, e.g., the rationalization of collections through regional and international collaboration, an improved data and information management, as well as the reduced over-duplication of samples. Furthermore, legally binding mechanisms should allow countries to place materials in secure storage facilities outside of their borders, without compromising their sovereign rights over such material[104].

Further cooperation is needed to increase the *effectiveness of PGRFA conservation*. Cooperation between the public and the private sector, the public conservation facilities and the private seed industry, as well as between the professional breeding and farmers in marginalized areas are recommended. The public sector can help the private sector in undertaking plant breeding, seed production, and distribution, by creating an enabling environment, including the appropriate legislation. The private sector requires certain prerequisites for investment; for instance, a base level of market infrastructure and

---

[103] This dilemma occurs for developing countries in most other areas of raw material export as well.

[104] For more detailed recommendations see the Global Plan of Action for the Conservation and Sustainable Utilization of Plant Genetic Resources for Food and Agriculture (FAO, 1996b).

communications, a reduction in restrictive regulations, the availability of capital and property protection.

The *breeding industry needs support* from public sector investment, particularly for the information processing and prebreeding activities, which represent long-term and high-risk programs of basic research especially in countries with an emerging breeding industry, like India, but also in countries with a majority of medium- and small-scale breeding companies, like Germany. These programs compete for resources with other long-term, but basic and applied research in the private breeding sector (Smith and Salhuana, 1996). The private investments are safer and higher when utilizing elite germplasm because of higher risks of failing when utilizing exotic germplasm and because of time constraints for the production of new varieties. Aside from increasing the attractiveness of PGRFA in genebanks by carrying out information processing and prebreeding, this task - carried out by genebanks and financed from the public sector - can be seen as support for all (private, public, and informal) breeding efforts. There are already some collaborative efforts among private seed industry with national and international public conservation facilities to support regeneration and evaluation, providing seed and information[105]. More of these collaborations between breeders and genebank managers will eventually lead to a more intensive use of landraces in breeding programs.

Furthermore, the breeding industry could cooperate with farmers in marginal areas. While the private sector's influence is currently rudimentary in marginal areas, because of poor infrastructure or the site-specific needs are of less interest to the private sector, cooperation with the individual breeding of farmers in marginal areas could, however, produce outputs of some interest for the breeding sector in the long run. Consequently, the breeding sector should support these farmers with seed and other inputs.

The final cooperation recommended is institutional linkages. As has been shown in the Indian and German case study, in situ conservation often involves institutions (ministries responsible for forestry and environment) other than those which have prime responsibility for ex situ conservation (the Ministry of Agriculture). Additionally, NGOs often play an important role in in situ management. These conservation efforts, however, are rarely coordinated between different "informal" activities as well as between NGOs' and public activities. Therefore effective co-ordination is necessary, thereby strengthening linkages between all formal and informal institutions and their ex situ and in situ conservation efforts. The " conservation and service centers" could take over the responsibility for such institutional linkages.

---

[105] E.g., Latin American Maize Project (LAMP): its objective is to conserve and evaluate maize landraces in a five year project, financially supported by the private seed industry (Pioneer Hi-Bred International, Inc.) in cooperation with Argentina, Bolivia, Brazil, Colombia, Chile, Guatemala, Mexico, Paraguay, Peru, Uruguay, Venezuela and USA (Smith and Salhuana, 1996), or the financial, human, and field assistance for the regeneration of CIMMYT's maize accessions (Smith and Salhuana, 1996).

8 Summary

Incomplete information and the uncertainty of the effects of PGRFA loss on global welfare as well as the uncertainty of the conservation costs hampers an economically efficient approach to diversity conservation. This study wants to contribute to the development and the conceptualizing of more efficient conservation strategies. These are preconditions for effective utilization of PGRFA for food security and a productive world agriculture in the long run.

Because of its generally assumed decline, the call for conservation of biodiversity and agrobiodiversity is implemented by a vast number of conservation activities. Although a decline in numbers of species (biodiversity) and crop varieties (agrobiodiversity) has been ascertained, the extinction of genetic diversity in agriculturally relevant crop varieties is not based on reliable estimations at present. The determinants of the loss of diversity and the underlying economic causes are highlighted. Although human-induced habitat modification, fragmentation, and destruction determine the decline in biodiversity, human activities were the precondition for the rise of agrobiodiversity through breeding as well as for its decline due to the abandonment of landraces and old varieties. In addition to socio-economic factors, the lack of adequate and sufficient policies, institutional arrangements, and markets for genetic resources hinder the internalization of the existing and expected social values of genetic resources.

Ex situ conservation, defined as storing PGRFA outside their natural or traditional ecosystems, has comparative advantages in realizing two of the three conservation objectives: freezing PGRFA for future use and improving the accessibility of conserved genetic resources. The in situ conservation, defined as the conservation of PGRFA in the locations in which the material has evolved, is, however, necessary for further adaptation of PGRFA to changing environmental conditions.

The call to conserve the maximum of genetic diversity is based on its valuation from a purely environmental-ethical point of view. In contrast, the economic valuation of PGRFA based on anthropocentric grounds encounters valuation problems, because of the intergenerational existence and the problem of intermittent scarcity and irreversible loss. Consequently, conventional instruments are of limited use when it comes to assessing genetic resources.

No remuneration has been offered so far; either for farmers who are maintaining PGRFA diversity in their fields ("in situ conservation") or for the ex situ conservators who are involved in the first processing of genetically coded information derived from PGRFA. The conservation system and an optimal level of agrobiodiversity will, however, only be sustained if the incentives are sufficient. Hence, the institutional aspects of an exchange system for PGRFA will determine the future conservation.

The institutional framework for the exchange of PGRFA between supply and demand is analyzed. In addition to the actors of conservation and exchange activities and their divergent interests, the development of an institutional framework is determined by the incentive system, to be based either on the principle of compensation or on the principle of internalization. This debate takes place between agrobiodiversity-rich and diversity-poor countries as well as between technologically rich and poor countries, both agrobiodiversity-rich. This discussion can be seen as the first step towards to a market or bilateral exchange system, whereas the existing informal multi-lateral exchange system will most likely evolve into a formal exchange system.

The demand for stronger property protection rights in the developing countries by the private sector and industrialized countries will affect the free access to PGRFA as well. The property rights will enable the agrobiodiversity-rich countries to derive additional benefits from their genetic resources. Furthermore, property rights protection will boost the development of the breeding sector, thereby enabling agrobiodiversity-rich countries to benefit from the additionally incorporated value of genetic resources, if those countries have a potential in breeding technologies.

This study analyzes the costs of conservation of plant genetic resources for food and agriculture (PGRFA), it assesses the conservation effectiveness, and the efficiency of the different conservation instruments. The database assembled and used for the empirical work consists of the costs at the level of storage facilities (e.g., genebanks) as well as the costs for conservation programs at the national and international level. The data utilized were extracted from a review published and unpublished sources, personal interviews with people in charge of some storage facilities, and from a questionnaire sent to all countries involved in the international conservation process.

The average ex situ conservation cost for one accession can presently be calculated with US $ 44. Costs vary significantly, however, because of the different conservation methods and sites. In comparing the ex situ conservation costs with those of in situ conservation programs as the "on-farm management" of approximately US $ 500 per conserved landrace, it is suggested to promote such programs only cautiously; especially as long as most of PGRFA still in situ are conserved by farmers without any conservation costs. Consequently, a system of competition is needed in which the conservation of landraces is safeguarded with some institutional framework. A suggested "controlled in situ conservation" system is based on a calculated areal minimum safety standard of less than 5 hectares for every landrace threatened by extinction. Consequently, it seems to be a more cost-effective way of introducing the necessary in situ conservation.

In analyzing the costs of some representative countries, the results show that some agrobiodiversity-rich but low-income countries spent as much for the conservation management as some agrobiodiversity-poor but high-income countries for the year 1995, if the expenditures were set into relation to the country's average per capita income. The governments from the latter group see a need to safeguard their breeding industry's demand for genetic resources as input for breeding. On the supply side, agrobiodiversity-rich countries invest much for the conservation of PGRFA, too. They do so for their own country's breeding efforts but also to be able to appear on a market to be developed for genetically coded information.

Based on expenditure data of 39 countries, the overall expenditures at a national as well as an international level were estimated for 1995. According to the estimation, US $ 733 million were spent for domestic expenditures in PGRFA conservation management. US $ 83 million were estimated to be contributed from countries for international organizations and their PGRFA conservation projects. It is estimated that at least US $ 106 million were contributed directly from country to country on a bi-lateral or multi-lateral basis.

Finally, the effectiveness of conservation in specific countries was analyzed. It is interesting to note that only five of 34 countries received a high rating for their effectiveness of PGRFA conservation. These countries were: Ethiopia, Germany, Japan, Poland, and the USA.

Following the detailed cost analysis, some conclusions for a practicable PGRFA conservation policy are drawn:

- "Conservation and service centers" should be established to promote the utilization of the stored PGRFA.
- Incentives for in situ conservation are necessary, but only in the form of site-specific contribution.
- Compensation and internalization fit with the political call for "benefit sharing". However, a full internalization of the social benefits at the individual farmer's level is presently impossible and not feasible at national level.
- Increasing effectiveness of conservation strategies can be improved through a market system, technology development, and the cooperation at different levels.

9 Zusammenfassung

Erhaltung pflanzengenetischer Ressourcen für Ernährung und Landwirtschaft: Kosten und Implikationen für eine nachhaltige Nutzung

Sowohl unvollständige Information und Ungewißheit der Wirkungen von Verlust der pflanzengenetischen Ressourcen für die Ernährung und Landwirtschaft (PGRFA) auf die globale Wohlfahrt als auch Ungewißheit der Erhaltungskosten beeinträchtigen einen ökonomisch effizienten Ansatz der PGRFA Erhaltung. Diese Arbeit soll einen Beitrag dazu leisten, in Zukunft effizientere Erhaltungsstrategien zu konzipieren und zu realisieren, denn PGRFA Erhaltung ist Voraussetzung für ihre Nutzung zur langfristigen Ernährungssicherung und für eine produktive Weltlandwirtschaft.

Der Forderung nach der Erhaltung der Biodiversität und der Agrobiodiversität wird durch eine Vielzahl von Maßnahmen zur Konservierung von PGRFA nachgekommen und mit dem allgemein angenommenen Verlust von genetischen Ressourcen begründet. Auch wenn ein Rückgang der Artenvielfalt (generelle Biodiversität) und der Sorten von Kulturpflanzen (Agrobiodiversität) festgestellt werden kann, so herrscht bisher Unklarheit über den Verlust von genetischer Vielfalt in den Kulturpflanzen. Die Determinanten des Verlustes der Artenvielfalt und der zugrundeliegenden wirtschaftlichen Ursachen werden hervorgehoben. Während menschlich induzierte Habitatveränderung, -fragmentierung und –zerstörung die wesentlichen Ursachen für den Rückgang der generellen Biodiversität darstellen, sind menschliche Aktivitäten Grund sowohl für die Entfaltung der Agrobiodiversität durch züchterische Tätigkeit als auch für deren Verlust durch die Aufgabe von Landsorten. Neben den sozioökonomischen Faktoren verhindern das Fehlen von angemessenen und hinreichenden Politiken, institutionellen Rahmenrichtlinien und Märkten für genetische Ressourcen die Internalisierung der bestehenden und erwarteten gesellschaftlichen Nutzen von genetischen Ressourcen.

Ex situ Erhaltung, definiert als die Erhaltung von PGRFA außerhalb ihres natürlichen Umfeldes, weist in zwei von drei Hauptzielen bei der PGRFA-Erhaltung komparative Vorteile auf: die Erhaltung von PGRFA für zukünftige Nutzung wie auch die Verbesserung des Zuganges zu konservierten genetischen Ressourcen. Allerdings ist die in situ Konservierung, definiert als die Erhaltung genetischer Ressourcen im natürlichen Lebensraum, notwendig für weitere Anpassung der Ressourcen an sich veränderte Umweltbedingungen.

Der Ruf nach einer vollständigen Erhaltung aller genetischer Ressourcen basiert vorwiegend auf einem umwelt-ethischen Standpunkt. Aufgrund der intergenerativen Werte pflanzengenetischer Ressourcen, sowie den Problemen der Verknappung und des irreversiblen Verlustes stößt die anthropozentrisch basierte ökonomische Bewertung auf Probleme. Daher haben konventionelle Bewertungsinstrumente bei der Bewertung von PGRFA eine beschränkte Verwendung.

Bisher wird keine Entschädigung denjenigen angeboten, die vorrangig die PGRFA erhalten. Weder die Landwirte, die die Agrobiodiversität in ihren Feldern erhalten (die „in situ Konservierung") noch die ex situ Erhalter, die die ersten Verarbeitungsschritte vom Rohprodukt hin zu der aufbereiteten genetisch kodierten Information durchführen, erhalten eine Entschädigung.

Jedoch kann sowohl das Erhaltungssystem als auch das optimale Niveau der Agrobiodiversität nur dann erhalten werden, wenn ausreichende Anreize entweder in Form von einer Internalisierung oder einer Kompensation angeboten werden. Die institutionellen Aspekte eines Austauschsystems für PGRFA werden daher die künftige Erhaltung bestimmen. Der institutionelle Rahmen für den Austausch von PGRFA zwischen Angebot und Nachfrage wird untersucht. Neben den Akteuren der Erhaltungs- und Austauschmaßnahmen und ihren divergierenden Interessen, wird die Entwicklung des institutionellen Rahmens durch das Anreizsystem bestimmt. Dieses kann auf dem Prinzip der Entschädigung oder der Internalisierung basieren. Die Diskussion um das richtige Anreizsystem findet sowohl zwischen Ländern mit hoher Agrobiodiversität und solchen mit geringer Agrobiodiversität wie auch zwischen Ländern mit einer hohen Agrobiodiversität aber unterschiedlicher Ausstattung an technologischem Potential statt. Diese Diskussion kann als der erste Schritt zu einem Austauschsystem angesehen werden, welches auf marktwirtschaftlichen Grundsätzen oder bilateralen Austauschmechanismen basiert.

Die Forderung des privaten Sektors und der Industrieländer nach verbesserten Eigentums-Schutzrechten in den Entwicklungsländern wird Auswirkungen auf den freien Zugang zu PGRFA haben. Auch wenn Länder mit einer hohen Agrobiodiversität immer noch skeptisch gegenüber verschärftem Rechtsschutz sind, werden Eigentumsrechte den Schutz ihrer genetischen Ressourcen verbessern. Zusätzlich fördern verbesserte Eigentumsrechte die Entwicklung des Zuchtsektors für solche Länder, die ein Technologiepotential aufzuweisen haben. Verschärfte Eigentumsrechte ermöglichen ihnen somit, den zusätzlichen Wert durch die Veredlung der genetischen Ressourcen selber abzuschöpfen.

Die vorliegende Arbeit analysiert die Kosten der Erhaltung der PGRFA und den Einfluß der Kosten sowohl auf die Wirksamkeit der Erhaltung als auch auf die Effizienz der verschiedenen Instrumente der Erhaltung. Als Datengrundlage für die empirische Analyse dienten einzelwirtschaftliche Kosten von unterschiedlichen Erhaltungseinrichtungen sowie von Erhaltungsprogrammen auf nationaler wie auch internationaler Ebene. Die Kosten wurden aus der Literatur, von Informationen einiger Erhaltungseinrichtungen und anhand eines Fragebogens ermittelt.

Die Durchschnittskosten für die ex situ Konservierung eines Musters (Accession) belaufen sich derzeit auf US $ 44. Allerdings variieren diese Kosten bedingt durch Konservierungsmethode und -ort. Der Vergleich mit den Kosten von in situ Konservierungsprogrammen, den sogenannten „on-farm management“ Aktivitäten, von ca. US $ 500 je erhaltener Landsorte, führt zu dem Schluß, daß solche in situ Konservierungsprogramme nur eingeschränkt gefördert werden sollten. Dieses um so mehr, als der vorwiegende Teil von PGRFA immer noch von Landwirten ohne jegliche Konservierungskosten in situ erhalten wird. Ein System mit freiem Wettbewerb zwischen Landsorten und modernen Sorten wird benötigt. Dabei muß jedoch die Erhaltung der Landsorten durch einen institutionellen Rahmen gewährleistet sein. Ein vorgestelltes System der „kontrollierten in situ Erhaltung“ basiert auf einem flächenbezogenen minimalen Sicherheitsstandard von weniger als 5 ha pro Landsorte, die vom Aussterben bedroht ist.

Weiterhin wurden die Erhaltungskosten einiger repräsentativer Länder für das Jahr 1995 analysiert. Die Ergebnisse deuten darauf hin, daß agrobiodiverse aber einkommensschwache Länder relativ ähnlich hohe Erhaltungsausgaben aufweisen wie einkommensstarke Länder mit einer geringen Agrobiodiversität, wenn die Ausgaben in Relation zum Durchschnittseinkommen gesetzt werden. Die Regierungen der letztgenannten Gruppe sehen die Notwendigkeit der (ex situ) Erhaltung genetischer Ressourcen, um die Nachfrage nach genetischen Ressourcen für die Zucht zu sichern. Die relativ hohen Ausgaben für die Erhaltung pflanzengenetischer Ressourcen der Entwicklungsländer mit hoher Agrobiodiversität werden einerseits bestimmt durch die Förderung der nationalen Zuchtbemühungen (des öffentlichen aber auch privaten Sektors), und andererseits durch die Bemühungen auf einem zukünftigen Markt für genetisch kodierten Informationen als Anbieter auftreten zu können.

Die Gesamtausgaben zur Erhaltung von PGRFA auf nationaler wie auch internationaler Ebene wurden für 1995 geschätzt, basierend auf Hochrechnungen mit Daten von 39 Ländern. Geschätzte US $ 733 Millionen wurden für Erhaltungsmaßnahmen innerhalb aller Länder investiert. Weitere US $ 82 Millionen wurden von Ländern für die Erhaltung von PGRFA durch internationale Organisationen und deren Projekten zur Verfügung gestellt. Auf der Basis von bilateralen oder multilateralen Abkommen sind mindestens geschätzte US $ 107 Millionen für Erhaltungsmaßnahmen zur Verfügung gestellt worden.

Abschließend wurde die Wirksamkeit von Erhaltungsmaßnahmen für 34 Länder untersucht. In lediglich fünf dieser Länder konnte eine hohe Wirksamkeit der Erhaltungsmaßnahmen nachgewiesen werden. Es sind: Äthiopien, Deutschland, Japan, Polen und die USA.

Schlußfolgerungen für eine praktische Politik der Erhaltung von PGRFA sind u.a.:

- Um die Nutzung der konservierten genetischen Ressourcen zu fördern, wird die Entwicklung von nationalen „Konservierungs- und Dienstleistungszentren“ vorgeschlagen.
- Anreize für die in situ Erhaltung von PGRFA sind notwendig, jedoch sollten diese nur bei akuter Gefahr des Aussterbens einer Landsorte eingesetzt werden.

- Beide Systeme, das der Kompensation wie auch der Internalisierung sind tauglich für die politische Forderung nach „benefit sharing", jedoch ist eine vollständige Internalisierung der sozialen Nutzen auf der individuellen Ebene des einzelnen Landwirts unmöglich und derzeit auf nationaler Stufe auch nicht durchführbar.
- Die Wirksamkeit der Erhaltungsstrategien kann durch den Wettbewerb zwischen Landsorten und modernen Sorten, durch die nationale Technologieentwicklung und durch die Zusammenarbeit auf verschiedenen Ebenen verbessert werden. Modelle für solche Kooperationen werden umrissen.

10 Appendices

Appendix 1: Share of wheat, rice, and maize accessions stored ex situ

	Accessions	Worldwide stored accessions [%]
Wheat (W):	850,145	14%
Rice (R):	420,341	7%
Maize (M):	261,584	4%
Total of W, R, M:	1,532,070	25%
Other crops:	4,615,802	75%
World Total:	6,147,872	

Source: calculated according to FAO, 1996a, WIEWS, 1996

Appendix 2: Share of CIMMYT's and IRRI's accessions stored ex situ

	Accessions	Total CGIAR stored accessions [%]
CIMMYT:	136,637	23%
IRRI:	80,646	14%
Other centers:	376,084	63%
CGIAR Total:	593,367	

Source: calculated according to SGRP, 1996

Appendix 3: Examples of the decline in varietal diversity of PGRFA

Country	Reduction of cultivated area for local varieties	Complete replacement of local varieties
Argentina	Amaranthus, quinoa	Amaranthus, quinoa
Chile		Varieties of potato, oats, barley, lentils, watermelon, tomato and wheat
China[d]		Of 10,000 wheat varieties used in 1949 only 1,000 were still in use by the 1970s 1975 100% of the rice-area is planted to modern varieties
Costa Rica	Varieties of maize and phaseolus vulgaris	
Ethiopia	Barley varieties	Durum wheat
India[a]	From 30,000 varieties to 75% of production from less than 10 varieties	
Malaysia, Philippines, Thailand[b]	Rice, maize, and fruit varieties	
Mexico	Only 20% of the local varieties reported in 1930 are now known in Mexico	
Republic of Korea		74% of varieties of 14 crops being grown on particular farms in 1985 had been replaced by 1993
Sri Lanka[b]	From 2,000 rice varieties in 1959 to 5 major varieties today	
Uruguay		Wheat varieties
USA[c]		Of the 7,098 apple varieties documented in the US Department of Agriculture as having been in use between 1804 and 1904, approximately 86% have been lost 95% of the cabbage varieties 91% of the field maize varieties 94% of the pea varieties 81% of the tomato varieties

Source: Country Reports of Argentina, Chile; China, Costa Rica, Ethiopia, Mexico, Republic of Korea, Uruguay,
[a]: FAO, 1993a; [b]: FAO, 1996a; [c]: Fowler, 1994; [d]: Kush, 1996

Appendix 4: Valuation of natural resources by different approaches

Valuation approach	Objective of valuation	Reference
Travel cost model (TCM)	Valuation of forests in the UK	Garrod and Wills, 1991
TCM	Value of achray forest in Central Scotland	Hanley, 1989
TCM	Measure of recreational fishing losses due to hydroelectric developments	Loomis et al., 1986
Contingent valuation method (CVM)	Value of preserving elephants in Kenya	Brown, 1993
CVM	Loggerhead sea turtle program	Whitehead, 1993
CVM	Exxon Valdez oil spill case	Carson et al., 1992
CVM	Value of preserving the northern spotted owl	Rubin et al., 1991
CVM	Value of preserving of striped shiner, Wisconsin	Boyle and Bishop, 1987
CVM	Existence value of eagles, Wisconsin	Boyle and Bishop, 1986
CVM	Value of blue whales, bottlenose dolphins, California sea otters, and northern elephant seals	Hagemann, 1985
CVM	Whooping crane population at the Arkansas National Wildlife Refuge in Texas	Stoll et Johnson, 1984
Production function approach (PFA)	Value of soil conservation	Pimentel et al., 1994
PFA	Value of coastal wetland	Aylward and Barbier, 1992
PFA	Value of mangrove system, Bintuni Bay, Irian Jaya, Indonesia	Ruitenbeek, 1992
PFA	Value of Hadejia-Jama'are floodplain in Northern Nigeria	Barbier et al., 1991
PFA	Calculating the social value of the marginal product of wetland areas	Freeman, 1991
PFA	Value of coral diversity	Hodgson and Dixon, 1988

Appendix 5: PGRFA conserved ex situ in OECD countries

Europe		America		Asia / Pacific		Total
Country	Number of accessions	Country	Number of accessions	Country	Number of accessions	
Austria	7,891	Canada	212,061	New Zealand	70,000	
Belgium	9,750	USA	550,000	Australia	123,200	
Czech Republic	51,571	Mexico	74,087	Japan	202,581	
Denmark	3,660					
Finland	2,323					
France	249,389					
Germany	200,000					
Greece	17,556					
Iceland	0					
Ireland	2,758					
Italy	80,000					
Luxembourg	0					
Netherlands	67,374					
Norway	1,133					
Portugal	29,361					
Spain	78,174					
Sweden	89,206					
Switzerland	17,000					
Turkey	40,000					
UK	114,495					
Total	1,021,641		836,148		202,581	2,060,370

Source: calculated according to information from FAO, 1996a, WIEWS database, 1996

Appendix 6: Ex situ conserved PGRFA by storage type

Storage type	Number of accessions
Seedbanks[a]	3,610,428
Fieldbanks[a]	526,300
In vitro[b]	37,600
Unknown storage facilities[c]	1,380,177
Total:	5,554,505

Source: [a] FAO, 1996a, [b] FAO, 1995d, [c] calculations according to information from WIEWS, 1996

Appendix 7: Estimated costs for in vitro conservation in developing countries

	Total annual average costs [US $]
Labor:	11,458.41
All other costs:	24,344.71
Total average fixed costs:	35,803.12
Annual average fixed costs for one accessions (by 491 accessions):	72.92
Total annual variable costs:	10.00
Annual average total costs for conservation preparation for one accessions:	**82.92**

Source: recalculated Gehl's data (1997a), taking average labor costs of 30% of European level

Appendix 8: Cryopreservation for potatoes at BAZ, Braunschweig

	Total annual average costs in US $	**Annual average costs for conservation preparation (once every 50 years)** [a] **in US $**	**Annual storage costs in US $**
Labor:	1,160.09	770.68	389.41
All other costs:	3,291.48	162.66	3,128.82
Total average fixed costs:	4,451.58	933.34	3,518.24
Annual average fixed costs for one accessions:	22.26		
Total annual variable costs:	0.04		
Annual average total costs for conservation preparation for one accessions:	22.29		

[a] The costs for the conservation preparation of each accession are incurred every 50 years, before storing the accession after collection or regeneration. Consequently, the costs are divided over the expected length of conservation for calculation purposes (which is at least 50 years).

Source: recalculated according to information from Gehl (1997b)

Appendix 9: Estimated costs for cryopreservation in developing countries

	Total annual average costs [US $]	Annual average costs for conservation preparation (once every 50 years)[b] [US $]	Annual storage osts [US $]
Labor:	348.03	231.20	116.82
All other costs:	3,291.48	162.66	3,128.82
Total average fixed costs:	3,639.51	393.86	3,245.64
Annual average fixed costs for one accessions:	18.20		
Total annual variable costs:	0.04		
Annual average total costs for conservation preparation for one accessions:	18.23		

Recalculated Appendix 8, taking average labor costs of 30% of European level; The costs for the conservation preparation of each accession are incurred every 50 years, before storing the accession after collection or regeneration. Consequently, the costs are divided over the expected length of conservation for calculation purposes (which is at least 50 years).
Source: recalculated according to information from Gehl (1997b)

Appendix 10: PGRFA impact of CGIAR's funding allocations

Program		Total [US $ '000 000]	PGRFA impact: [%]	Total PGRFA [US $ '000 000]
1.1	Germplasm enhancement and breeding:	57.1	30	17.13
2	Protecting environment (incl. research on conservation of natural vegetation)	48.2	5	2.41
3	Saving biodiversity	27.2	100	27.20
5	Fortifying NARS	53.0	5	2.65
	Research program implementation and design	11.7	5	0.59
	Total			49.98

Source: adapted according to data from CGIAR, 1996

Appendix 11: Assumed distribution of ex situ conserved PGRFA according to storage type

Storage type	Number of accessions	Distribution [%]
Seed genebanks	3,610,428	65%
Field genebanks	526,300	9%
In vitro	37,600	1%
Unknown storage facilities	1,380,177	25%
Total:	5,554,505	100%

Source: calculated according to information from FAO, 1996a, WIEWS, 1996

The storage facilities are unknown for 25% of all conserved accessions, but it may be assumed that:

- 90% of these accessions are stored in seed genebanks;
- 8% of these accessions are stored in field genebanks; and
- 2% of these accessions are stored in in vitro storage facilities.

Consequently, the overall distribution of accessions is listed according to storage type:

Storage Type	Known storage distribution [%]	Assumed distribution of unknown storage facilities [%]	Assumed overall distribution of ex situ conserved accessions according to storage type [%]
Seed genebank	65.0%	22.5%	88.0%
Field genebank	9.0%	2.0%	11.0%
In vitro	1.0%	0.5%	1.0%
Cryopreservation	0.2%		0.2%

Appendix 12: Countries included and excluded in the estimation of conservation expenditures in 1995

Countries included	Countries excluded
Afghanistan, Albania, Algeria, Angola, Antigua and Barbuda, Argentina, Armenia, Australia, Austria, Azerbaijan, Bahamas, The, Bangladesh, Barbados, Belarus, Belgium, Belize, Benin, Bhutan, Bolivia, Botswana, Brazil, Bulgaria, Burkina Faso, Burundi, Cambodia, Cameroon, Canada, Cape Verde, Central African Republic, Chad, Chile, China, Colombia, Congo, Costa Rica, Cote d'Ivoire, Croatia, Cuba, Cyprus, Czech Republic, Denmark, Dominica, Dominican Republic, Ecuador, Egypt, Arab Republic of, El Salvador, Equatorial Guinea, Eritrea, Estonia, Ethiopia, Fiji, Finland, France, Gabon, Gambia, The, Germany, Ghana, Greece, Grenada, Guatemala, Guinea, Guinea-Bissau, Guyana, Haiti, Honduras, Hungary, Iceland, India, Indonesia, Iran, Islamic Republic of, Iraq, Ireland, Israel, Italy, Jamaica, Japan, Jordan, Kazakhstan, Kenya, Kiribati, Korea, Republic of, Latvia, Lebanon, Lesotho, Liberia, Libya, Lithuania, Madagascar, Malawi, Malaysia, Maldives, Mali, Mauritania, Mauritius, Mexico, Moldavia, Mongolia, Morocco, Mozambique, Myanmar, Namibia, Nepal, Netherlands, New Zealand, Nicaragua, Niger, Nigeria, Norway, Oman, Pakistan, Panama, Papua New Guinea, Paraguay, Peru, Philippines, Poland, Portugal, Puerto Rico, Romania, Russian Federation, Rwanda, Sao Tome and Principe, Saudi Arabia, Senegal, Seychelles, Sierra Leone, Slovak Republic, Slovenia, Solomon Islands, Somalia, South Africa, Spain, Sri Lanka, St. Kitts and Nevis, St. Lucia, St. Vincent and the Grenadines, Sudan, Suriname, Swaziland, Sweden, Switzerland, Syrian Arab Republic, Tanzania, Thailand, Togo, Tonga, Trinidad and Tobago, Tunisia, Turkey, Turkmenistan, Uganda, Ukraine, United Kingdom, United States, Uruguay, Uzbekistan, Vanuatu, Venezuela, Viet Nam, Western Samoa, Yemen, Republic of, Yugoslavia, Federal Republic of, Zaire, Zambia, Zimbabwe	Bahrain, Bosnia and Herzegovina, British Virgin Is., Brunei, Comoros, Djibouti, French Guyana, Georgia, Guadeloupe, Democratic People's Republic of Korea, Kuwait, Kyrgystan, Laos, Liechtenstein, Luxembourg, Macedonian, Malta, Marshall Islands, Micronesia, Monaco, Monserrat, Nauru, Palau, Palestine, San Marino, Singapore, Tajikistan, Tuiks and Caicos, Tuvalu, United Arab Emirates

Appendix 13: Domestic Expenditures 1995 for PGRFA Conservation

Countries	Domestic expenditures 1995 incl. foreign assistance [US $ '000]	Domestic expenditures - GDP/capita ratio
India	6,776	20.57
Egypt	11,528	14.68
Ethiopia	1,346	13.75
South Africa	19,000	6.06
China	2,526	5.78
Germany	113,215	4.77
France	98,660	4.52
UK	70,154	4.31
Spain	33,413	2.73
Tanzania	187	2.04
Brazil	8,000	1.92
Peru	4,137	1.84
Italy	27,208	1.59
Slovak Republic	3,608	1.55
Greece	10,958	1.54
Czech Republic	3,255	0.93
Russia	1,526	0.83
USA	20,433	0.83
Togo	151	0.61
Romania	408	0.32
Pakistan	120	0.27
Poland	656	0.26
Japan	6,480	0.19
Portugal	1,030	0.12
Switzerland	3,825	0.11
Canada	1,584	0.08
Austria	10	0.004

Appendix 14: Approach to estimation calculations

GDP: GDP is taken as criterion the first approach. The overall expenditures were estimated by correlating the overall GDP of all 165 countries to the sum of the specific countries' GDP and multiplying it with the known expenditures (for financial assistance received, contributed, and domestic expenditures respectively). The overall expenditures for the conservation of PGRFA, extrapolated for 165 countries, can be calculated as follows:

$$\sum_{n=1}^{165} Exp_x = \frac{\sum_{n=1}^{165} GDP}{\sum_{n=1}^{x} GDP} \times \sum_{n=1}^{x} Exp_x \qquad \text{(A.1)}$$

whereby:

$\Sigma\, Exp_x$ is the overall extrapolated expenditures for all countries (n=1...165) in one of the three categories (financial assistance received, financial assistance contributed, and domestic expenditures);

x defines the countries eligible for the relevant category (financial assistance received, financial assistance contributed, and domestic expenditures);

$\Sigma\, GDP$ is the total of GDP for all countries (n=1...165); or

$\Sigma\, GDP$ is the total of GDP for those countries only, belonging to the category calculated (n=1...x);

$\Sigma\, Exp_x$ is the sum of expenditures of the countries eligible (n=1...x).

Accessions: The ex situ conserved accessions were used as criterion in the second approach based on the assumption that the lion's share of national expenditures of the conservation of PGRFA is made in the area of ex situ conservation. The overall expenditures for the three categories can be extrapolated by calculating the ratio of all ex situ conserved accessions to the accessions conserved by the 39 countries and multiplying it with the known expenditures of these countries.

$$\sum_{n=1}^{165} Exp_x = \frac{\sum_{n=1}^{165} Acc}{\sum_{n=1}^{x} Acc} \times \sum_{n=1}^{x} Exp_x \qquad \text{(A.2)}$$

whereby:

$\Sigma\, Exp_x$ is the overall extrapolated expenditures for all countries (n=1...165) in one of the three categories (financial assistance received, financial assistance contributed, and domestic expenditures);

x defines the countries eligible for the relevant category (financial assistance received, financial assistance contributed, and domestic expenditures);

ΣAcc is the total of accessions conserved in all countries (n=1...165); or
ΣAcc is the total of accessions conserved in those countries only, belonging to the category calculated (n=1...x);
ΣExp_x is the sum of expenditures of the countries eligible (n=1...x).
The approach grouping the countries into 9 groups according to their stored accessions and according to their GDP/capita is presented as an example for the calculations can be seen as Appendix 20.

Appendix 15: Storage quality determined by regeneration requirement

Country	Accessions to be regenerated [%]	Quality standard of storage	Assumptions
Austria	10%	good	
Canada	n.i.	good	No data available, but due to 85% germination of seed for regeneration a good quality standard may be assumed.
China		good	China has not yet started to regenerate the conserved accessions, due to the date of establishing the gene bank (8 years ago), a good quality standard may be assumed.
Ethiopia	8%	good	
Japan	4%	good	
Poland	3%	good	
Switzerland	10%	good	
USA	19%	good	
Belarus		basic	Predominantly in vitro conservation with basic quality.
Cyprus	50%	basic	
Czech Republic	40%	basic	
France	n.i.	basic	No detailed information provided, concluding from Country Report's general information.
Germany	33%	basic	
Greece	50%	basic	
Ireland		basic	Sample collection began in 1994, after establishing a gene bank, but backlogs in regeneration started already.
Pakistan	48%	basic	
Peru		basic	No data available from Peru, but due to 50 to 60% germination of seed for regeneration a basic quality standard may be assumed.
Portugal		basic	Portugal has not yet started to regenerate the conserved accessions, but due to 70 to 80% germination of seed for regeneration a basic quality standard may be assumed.
Romania	45%	basic	
Russian Fed.		basic	Regeneration previously carried out by regional stations. With the disintegration of the USSR, the number of stations is reduced, increasing the regeneration problems, but for the moment, a basic quality may be assumed.
Slovak Rep.		basic	
United Kingdom		basic	No specific information given, but general storage conditions comply with international standards.

Appendix 15: Storage quality contd.

Brazil	64%	poor	
Egypt	100%	poor	Storage facilities are unreliable according to the CR.
Haiti		poor	Haiti has no ex situ conservation.
India	63%	poor	
Madagascar		poor	No information is provided, it may be assumed that the standard is more likely poor, esp. considering that the long-term stores seem to be unreliable.
Seychelles		poor	Because no information is provided it may be assumed that standard is more likely poor.
South Africa		poor	Some storage facilities are unreliable, for instance, some accessions have been lost due to lack of regeneration.
Spain		poor	No information is provided, it may be assumed that the standard is more likely poor.
Suriname		poor	Suriname has no ex situ conservation
Tanzania		poor	No information is provided, it may be assumed that the standard is more likely poor.
Togo		poor	No information is provided, it may be assumed that the standard is more likely poor.
Tonga		poor	All ex situ material is conserved as tissue culture, but because no information is provided it may be assumed that the standard is more likely poor

Storage quality is assessed for medium and short-term storage for countries without long-term storage facilities.
good quality: less than 20% regeneration requirements;
basic quality: more than 20% regeneration requirements;
poor quality: more than 50% regeneration requirements.
n.i. = no information
Source: respective Country Reports, FAO, 1996h

Appendix 16: Quality standard of documentation of storage facilities according to the existence of passport, characterization, and evaluation data

Country	Passport [%]	Characterization [%]	Evaluation [%]	Standard of documentation
Czech Republic	100	60	60	good
Ethiopia	100	100	100	good
France	100	90 [a]		good
Germany	100	90		good
Japan	100	100	100	good
Poland	100		68	good
United Kingdom	100	60 [a]	60 [a]	good
USA	100	60 [a]	60 [a]	good
Brazil	90	30 [a]	30 [a]	basic
Cyprus	100	30 [a]	30 [a]	basic
Greece	100	30 [a]	30 [a]	basic
Pakistan	80	60 [a]	60 [a]	basic
Peru	100	33	31 [a]	basic
South Africa				basic
Spain	100	60 [a]	30 [a]	basic
Canada[1]	100			basic
Austria	100			poor
Belarus	30 [a]	30 [a]	30 [a]	poor
China	95			poor
Egypt		15	15	poor
India				poor
Ireland	100			poor
Portugal	30 [a]	8	5	poor
Romania	80			poor
Russia	100			poor
Seychelles	20	5	90	poor
Slovak Republic			28	poor
Switzerland	100			poor
Tanzania				poor
Togo				poor
Tonga	No information at all			poor
Madagascar	No information at all			poor
Haiti	0	0	0	non-existent
Suriname				non-existent

[1] In Canada, characterization and evaluation is carried out by private breeders in interest of specific varieties, but not all data is sent back to the conservation facilities.

[a] In some Country Reports, the amount of documented accessions is only described with some, most or varies; these are interpreted for the effectiveness analysis as approximately 30%, 90%, and 60% respectively.

To group the documentation standard according to the documentation information,

- a good quality standard is given by more than 80% passport data and more than 50% other data;
- a basic quality standard is given by more than 80% passport data and between 30 and 50% other data;
- a poor quality standard is given by more than 80% passport data and more than 50% other data or even 100% passport but no other data.

Appendix 17: Distribution of accessions in 1994

Country	Distribution of accessions in 1994	Accessions according to CR	Distribution of PGRFA as % of national accessions	Standard of distribution of PGRFA
USA	127,539	550,000	23%	good
Ethiopia	6,000	54,000	11%	good
Germany	19,000	200,000	10%	good
Pakistan	1,000	18,000	6%	basic
Poland	3647	91,802	4%	basic
Canada	7,457	212,061	4%	basic
Czech Rep.	1,600	51,571	3%	basic
Japan	6,729	202,581	3%	basic
Brazil	1,264	194,000	1%	poor
Russia	2,827	333,727	1%	poor

If the distribution of germplasm in 1994 was 10% of the stored accessions or more, the standard of distribution was judged as good quality standard. A basic quality standard is defined by less than 10% but more than 3% of the stored accessions. If the conservation facilities distributed less than 3% of their stored material, their quality standard is defined as poor.

Source: respective Country Reports and WIEWS, 1996

Appendix 18: Underlying calculation for the assessment of the accessibility efforts

Based on the three discussed key indicators and sub-objectives, the countries' ex situ conservation facilities can be assessed concerning the accessibility of their germplasm. By developing a matrix, the single judgement for the three different criteria can be merged in one overall assessment. A good quality standard of documentation and distribution scored 3 points, a basic standard 2 and a poor standard 0.9 points. In addition, the existence of working collections in conservation facilities scores an additional point, whereas no point is given to the non-existence of a working collection:

Standard of documentation						Working Collection	
3	good	4.9	6	7	existent	1	
3		3.9	5	6	non-existent	0	
2	basic	3.9	5	6	existent	1	
2		2.9	4	5	non-existent	0	
0.9	poor	2.8	3.9	4.9	existent	1	
0.9		1.8	2.9	3.9	non-existent	0	
		poor	basic	good			
		Standard of distribution of PGRFA in 1994					

less than 3 represents a:	poor quality
between 3 and 5 represents a:	basic standard
over 5 represents a:	good standard

The overall quality standard for the accessibility of PGRFA germplasm conserved ex situ can be concluded from the calculations described. A good quality standard is achieved by a value over 5, basic standard is still possible over 3, and a poor standard is defined to be less than 3. In this way, all the surveyed countries can be classified into one of the three quality standards as it was done in Table 5.18 in Chapter 5.4.

Appendix 19: Underlying calculation for the assessment of the total effectiveness of PGRFA conservation

Based on the same approach as described in Chapter 5.4, the total effectiveness of the ex situ conservation facilities for each country is based on the assessments of the three single analyses. These have to be merged into an overall assessment, which is based on the following assumptions and calculations and depicted as can be seen:

Freezing							Adaptation
good	3.9	5.9	6.9	7.9	1	existent	
	3.9	4.9	5.9	6.9	0	non-existent	
basic	2.6	4.6	5.6	6.6	1	existent	
	2.6	3.6	4.6	5.6	0	non-existent	
poor	1.3	3.3	4.3	5.3	1	existent	
	1.3	2.3	3.3	4.3	0	non-existent	
		poor	basic	good			
		1	2	3			
			Access				

Note:

whereby:	<5:	low effectiveness;	5 – 6:	medium effectiveness;
	> 6:	high effectiveness		

If a country's conservation activities – as regards the long-term conservation perspective - can be assessed to be of good quality; this partial effectiveness receives scores 3.9 points, it scores 2.6 and 1.3 points, if it is of basic or poor quality respectively.

The same scoring principle was applied to the second objective of conservation activities, assessing the accessibility of stored material. If the quality of accessibility was assessed as poor, the partial effectiveness of a specific country's conservation activities scores 1 point. If it is estimated to be basic, it scores 2 points. If it is assumed to be of good quality it scores 3 points. An additional point is scored if in situ conservation activities were existent and therefore the potential of the germplasm for adaptation.

As can be seen in Table 5.16 the rating for the partial effectiveness is higher for the long-term conservation objective (good quality scores 3.9 points, whereas for the accessibility, a good quality scores only 3 points). The theory behind this approach reflects the idea that a good long-term conservation quality is the major contribution to the overall goal achievement of conservation PGRFA. Consequently, the scores are rated higher by 30%.

The overall quality standard for the accessibility of PGRFA germplasm conserved ex situ can be concluded from the calculations described. A good quality standard is achieved by an value over 5, basic standard is still possible over 3, and a poor standard is defined to be less than 3. In this way, all the surveyed countries can be classified into one of the three quality standards as was done in Table 5.18 in Chapter 5.4.

Appendix 20: Estimation of expenditures according to accessions and GDP/capita

Group: low accessions and low GDP/c						
Countries	GDP at market prices	GDP/capita	Genebank accessions	Foreign aid received 1995	Foreign aid contributed 1995	Domestic expenditures 1995 incl. for.assist.
-	[US $ '000]	[US $]	-	[US $ '000]		
Angola	183,000	19	599	-	-	-
Somalia	335,000	40	94	-	-	-
Mozambique	1,467,193	89	1,872	-	-	-
Tanzania	2,373,164	91	2,510	-	-	187
Eritrea	487,865	139	1,087	-	-	-
Chad	909,943	152	69	-	-	-
Burundi	909,465	156	0	-	-	-
Bhutan	278,266	186	40	-	-	-
Sudan	4,953,000	187	5,178	-	-	-
Niger	1,540,166	189	0	-	-	-
Sierra Leone	833,713	191	1,848	-	-	-
Nepal	3,970,515	200	8,383	-	-	-
Rwanda	1,494,073	204	6,168	-	-	-
Mali	1,870,310	209	248	-	-	-
Guinea-Bissau	242,559	237	0	-	-	-
Sao Tome and Principe	28,937	239	0	-	-	-
Haiti	1,645,103	245	0	796	-	1,896
Cambodia	2,246,154	248	2,155	-	-	-
Togo	973,343	250	4,000	-	-	151
Burkina Faso	2,814,664	295	850	-	-	-
Benin	1,522,154	302	2,453	-	-	-
Central African Rep.	980,187	309	0	-	-	-
Yemen, Republic of	4,654,370	358	4,229	-	-	-
Turkmenistan	1,471,000	381	4,832	-	-	-
Gambia, The	378,584	383	0	-	-	-
Equatorial Guinea	178,175	408	0	-	-	-
Liberia	973,000	410	1,707	-	-	-
Zambia	3,481,428	421	5,901	-	-	-
Nicaragua	1,771,029	457	2,976	-	-	-

Group: low accessions and low GDP/c contd.						
Countries	GDP at market prices	GDP/capita	Genebank accessions	Foreign aid received 1995	Foreign aid contributed 1995	Domestic expenditures 1995 incl. for.assist.
-	[US $ '000]	[US $]	-	[US $ '000]		
Lesotho	886,278	478	0	172	-	615
Kiribati	36,724	490	14	-	-	-
Mauritania	1,027,117	494	0	-	-	-
Ghana	8,004,287	507	2,987	-	-	-
Guinea	3,501,792	574	899	-	-	-
Honduras	3,161,140	583	4,457	-	-	-
Cameroon	7,469,530	610	2,329	-	-	-
Suriname	251,662	623	0	778	-	1,028
Congo	1,577,504	650	1,755	-	-	-
Guyana	540,278	670	0	-	-	-
Solomon Islands	240,500	718	1,130	-	-	-
Egypt	42,923,082	785	8,914	3,551	244	11,528
Moldova	3,726,577	855	6,000	-	-	-
Cape Verde	333,596	858	0	-	-	-
Western Samoa	148,400	916	0	-	-	-
Vanuatu	180,748	1,159	664	-	-	-
Papua New Guinea	4,724,466	1,165	5,656	-	-	-
Maldives	267,145	1,167	0	-	-	-
Armenia	4,345,606	1,182	2,000	-	-	-
Swaziland	1,102,056	1,284	0	-	-	-
Guatemala	12,855,090	1,320	2,796	-	-	-
Dominican Republic	10,415,500	1,423	2,024	-	-	-
Myanmar	62,212,538	1,423	8,000	-	-	-
Tonga	135,899	1,477	8	26	-	56
Total [US '000]	215,033,875	28,405	106,832	5,323	244	15,461

Group: low accessions and low GDP/c contd.						
Countries	GDP at market prices	GDP/capita	Genebank accessions	Foreign aid received 1995	Foreign aid contributed 1995	Domestic expenditures 1995 incl. for.assist.
-	[US $ '000]	[US $]	-		[US $ '000]	
53 Countries						
Foreign aid received by countries with GDP total of [US $ '000]	45,842,023	Estimated foreign aid received [US $ '000]		24,971		
Foreign aid contributed by countries with a GDP total of [US $ '000]	42,926,082	Estimated foreign aid contributed [US $ '000]			1,220	
Domestic Expenditures of countries with a GDP total of [US $ '000]	49,188,530	Estimated domestic expenditures [US $ '000]				67,589

Group: low accessions and medium GDP/c						
Countries	GDP at market prices	GDP/capita	Genebank accessions	Foreign aid received 1995	Foreign aid contributed 1995	Domestic expenditures 1995 incl. for.assist.
-	[US $ '000]	[US $]	-		[US $ '000]	
Jordan	6,105,352	1,546	3,588	-	-	-
Algeria	42,305,819	1,611	985	-	-	-
El Salvador	8,805,740	1,637	1,547	-	-	-
Paraguay	7,751,655	1,715	1,571	-	-	-
Jamaica	4,230,675	1,766	795	-	-	-
Tunisia	15,770,490	1,873	1,768	-	-	-
Namibia	2,883,857	1,892	1,600	-	-	-
Afghanistan	41,845,000	1,943	2,965	-	-	-
Belarus	20,286,831	1,970	4,000	-	-	135
St. Vincent and the Grena-dine	234,741	2,156	0	-	-	-
Latvia	5,816,817	2,203	9,730	-	-	-
Grenada	221,704	2,436	0	-	-	-
Lebanon	9,239,509	2,444	0	-	-	-
Fiji	1,867,359	2,490	943	-	-	-
Syrian Arab Republic	33,050,000	2,551	8,750	-	-	-
Costa Rica	8,283,885	2,594	5,057	-	-	-
Panama	6,851,900	2,724	1,538	-	-	-
Dominica	196,185	2,729	0	-	-	-
Belize	550,100	2,759	80	-	-	-
Estonia	4,578,158	2,946	3,000	-	-	-
Botswana	4,011,303	2,947	3,390	-	-	-
Mauritius	3,361,012	3,058	3,310	-	-	-
St. Lucia	495,667	3,196	58	-	-	-
Gabon	3,945,425	3,285	91	-	-	-
Iraq	68,603,000	3,580	6,400	-	-	-
Trinidad and Tobago	4,794,539	3,781	2,315	-	-	-
St. Kitts and Nevis	194,963	4,688	0	-	-	20
Total in US [US $ '000]	306,281,684	68,522	63,481	-	-	155

Group: low accessions and medium GDP/c contd.						
Countries	GDP at market prices	GDP/capita	Genebank accessions	Foreign aid received 1995	Foreign aid contributed 1995	Domestic expenditures 1995 incl. for.assist.
-	[US $ '000]	[US $]	-		[US $ '000]	
27 Countries						
Foreign aid received by countries with GDP total of [US $ '000]	0	Estimated foreign aid received [US $ '000]		-		
Foreign aid contributed by countries with a GDP total of [US $ '000]	0	Estimated foreign aid contributed [US $ '000]			-	
Domestic Expenditures of countries with a GDP total of [US $ '000]	20,481,794	Estimated domestic expenditures [US $ '000]				2,318

Group: low accessions and high GDP/c						
Countries	GDP at market prices	GDP/capita	Genebank accessions	Foreign aid received 1995	Foreign aid contributed 1995	Domestic expenditures 1995 incl. for.assist.
-	[cur. US $ '000]	[US $]	-		[US $ '000]	
Uruguay	16,154,020	5,161	1,256	-	-	-
Libya	29,838,000	6,131	2,313	-	-	-
Barbados	1,631,035	6,300	2,868	-	-	-
Seychelles	442,192	6,390	369	302	-	2,322
Slovenia	12,825,250	6,416	2,676	-	-	-
Saudi Arabia	111,342,797	6,615	0	-	-	-
Antigua and Barbuda	457,111	6,929	0	-	-	-
Oman	11,628,090	7,056	238	-	-	-
Puerto Rico	35,833,602	10,009	4,000	-	-	-
Bahamas, The	3,065,000	11,698	0	-	-	-
Ireland	47,678,022	13,442	2,758	0	142	-
Finland	84,111,639	16,666	2,323	0	1,180	-
Belgium	210,576,392	20,974	9,750	0	-	-
Austria	182,067,102	23,096	7,891	0	1,500	10
Iceland	6,075,987	23,280	0	0	-	-
Norway	103,418,601	24,129	1,133	0	2,612	16,208
Denmark	134,677,103	26,050	3,660	0	-	-
Total [US '000]	991,821,942	220,342	41,235	302	5,434	18,540
17 Countries						
Foreign aid received by countries with GDP total of [US $ '000]	769,047,037	Estimated foreign aid received [US $ '000]		389		
Foreign aid contributed by countries with a GDP total of [US $ '000]	417,275,363	Estimated foreign aid contributed [US $ '000]			12,916	
Domestic Expenditures of countries with a GDP total of [US $ '000]	285,927,895	Estimated domestic expenditures [US $ '000]				64,311

Group: medium accessions and low GDP/c						
Countries	GDP at market prices	GDP/capita	Genebank accessions	Foreign aid received 1995	Foreign aid contributed 1995	Domestic expenditures 1995 incl. for.assist.
-	[cur. US $ '000]	[US $]	-		[US $ '000]	
Zaire	3,714,000	93	18,830	-	-	-
Malawi	1,301,862	144	11,421	-	-	-
Viet Nam	15,570,000	225	21,493	-	-	-
Madagascar	2,816,562	228	15,000	1,577	-	2,385
Bangladesh	26,163,470	229	45,309	-	-	-
Uganda	4,031,764	231	11,483	-	-	-
Mongolia	740,699	321	24,000	-	-	-
Nigeria	35,200,070	345	12,324	-	-	-
Pakistan	52,011,172	436	19,208	-	-	120
Azerbaijan	3,647,243	495	25,000	-	-	-
Senegal	3,881,485	495	12,000	-	-	-
Cote d'Ivoire	6,716,319	521	22,498	-	-	-
Zimbabwe	5,432,434	524	45,698	-	-	-
Albania	1,808,078	538	20,000	-	-	-
Sri Lanka	11,714,690	673	11,781	-	-	-
Bolivia	5,511,840	733	11,069	-	-	-
Indonesia	174,638,399	948	26,828	-	-	-
Total [US '000]	354,900,087	7,177	353,942	1,577	-	2,505
17 Countries						
Foreign aid received by countries with GDP total of [US $ '000]	2,816,562	Estimated foreign aid received [US $ '000]		198,759		
Foreign aid contributed by countries with a GDP total of [US $ '000]	0	Estimated foreign aid contributed [US $ '000]			-	
Domestic Expenditures of countries with a GDP total of [US $ '000]	54,827,734	Estimated domestic expenditure s [US $ '000]				16,214

Group: medium accessions and medium GDP/c						
Countries	GDP at market prices	GDP/capita	Genebank accessions	Foreign aid received 1995	Foreign aid contributed 1995	Domestic expenditures 1995 incl. for.assist.
-	[cur. US $ '000]	[US $]	-		[US $ '000]	
Kazakhstan	18,327,890	1,076	33,000	-	-	-
Morocco	31,503,200	1,203	20,470	-	-	-
Lithuania	5,223,750	1,391	12,821	-	-	-
Ecuador	16,557,370	1,502	35,780	-	-	-
Cuba	16,585,000	1,533	18,668	-	-	-
Iran, Islamic Republic of	113,273,504	1,900	40,000	-	-	-
Turkey	131,013,902	2,238	26,867	-	-	-
Peru	50,384,552	2,252	44,833	1,883	-	4,137
Slovak Republic	12,369,570	2,334	14,547	-	-	3,608
Thailand	143,208,792	2,469	32,404	-	-	-
Croatia	13,459,710	2,811	15,336	-	-	-
Venezuela	58,882,470	2,908	15,356	-	-	-
South Africa	124,726,198	3,137	48,918	-	-	19,000
Malaysia	70,634,160	3,796	38,255	-	-	-
Yugoslavia, Fed. Rep. of	40,463,000	3,818	38,000	-	-	-
Chile	52,180,918	3,837	36,000	-	-	-
Total [US $ '000]	898,793,987	38,205	471,255	1,883	-	26,745
16 Countries						
Foreign aid received by countries with GDP total of [US $ '000]	50,384,552	Estimated foreign aid received [US $ '000]		33,594		
Foreign aid contributed by countries with a GDP total of [US $ '000]	0	Estimated foreign aid contributed [US $ '000]			-	
Domestic Expenditures of countries with a GDP total of [US $ '000]	187,480,320	Estimated domestic expenditures [US $ '000]				128,217

Group: medium accessions and high GDP/c						
Countries	GDP at market prices	GDP/capita	Genebank accessions	Foreign aid received 1995	Foreign aid contributed 1995	Domestic expenditures 1995 incl. for.assist.
-	[US $ '000]	[US $]	-		[US $ '000]	
Greece	73,109,504	7,098	17,556	-	-	10,958
Argentina	281,922,109	8,517	30,000	-	-	-
Portugal	84,736,459	8,606	29,361	-	10	1,030
Cyprus	6,480,901	9,026	12,313	7	-	186
New Zealand	43,698,770	12,692	28,914	-	-	-
Switzerland	232,160,592	33,622	17,000	-	3,400	3,825
Total [US $ '000]	722,108,335	79,562	135,144	7	3,410	15,999
6 Countries						
Foreign aid received by countries with GDP total of [US $ '000]	6,480,901	Estimated foreign aid received [US $ '000]		780		
Foreign aid contributed by countries with a GDP total of [US $ '000]	316,897,051	Estimated foreign aid contributed [US $ '000]			7,770	
Domestic Expenditures of countries with a GDP total of [US $ '000]	396,487,456	Estimated domestic expenditures [US $ '000]				29,138

Group: high accessions and low GDP/c						
Countries	GDP at market prices	GDP/capita	Genebank accessions	Foreign aid received 1995	Foreign aid contributed 1995	Domestic expenditures 1995 incl. for.assist.
-	[US $ '000]	[US $]	-		[US $ '000]	
Ethiopia	5,366,250	98	54,000	969	-	1,346
Kenya	6,955,602	271	50,037	-	-	-
India	291,054,191	329	342,108	4,565	-	6,776
China	508,179,612	437	350,000	1,350	-	2,526
Uzbekistan	21,248,530	990	50,000	-	-	-
Philippines	63,876,121	994	59,399	-	-	-
Total [US $ '000]	896,680,305	3,120	905,544	6,884	-	10,648
6 Countries						
Foreign aid received by countries with GDP total of [US $ '000]	804,600,052	Estimated foreign aid received [US $ '000]		7,672		
Foreign aid contributed by countries with a GDP total of [US $ '000]	0	Estimated foreign aid contributed [US $ '000]				
Domestic Expenditures of countries with a GDP total of [US $ '000]	804,600,052	Estimated domestic expenditures [US $ '000]				11,867

Group: high accessions and medium GDP/c						
Countries	GDP at market prices	GDP/capita	Genebank accessions	Foreign aid received 1995	Foreign aid contributed 1995	Domestic expenditures 1995 incl. for.assist.
-	[cur. US $ '000]	[US $]	-		[US $ '000]	
Bulgaria	10,114,910	1,189	55,420	-	-	-
Romania	29,205,660	1,284	93,000	-	-	408
Ukraine	76,139,250	1,461	136,400	-	-	-
Russian Federation	274,929,091	1,845	333,000	260	-	1,526
Colombia	64,371,429	1,927	85,000	-	-	-
Poland	95,406,342	2,487	91,802	10	-	656
Czech Republic	36,024,300	3,492	51,571	-	-	3,255
Hungary	41,175,351	3,993	75,170	-	-	-
Brazil	639,562,416	4,157	194,000	1,112	-	8,000
Mexico	375,465,116	4,419	103,305	-	-	-
Total [US $ '000]	1,642,393,864	26,254	1,218,668	1,382	-	13,846
10 Countries						
Foreign aid received by countries with GDP total of [US $ '000]	1,009,897,849	Estimated foreign aid received [US $ '000]		2,248		
Foreign aid contributed by countries with a GDP total of [US $ '000]	0	Estimated foreign aid contributed [US $ '000]			-	
Domestic Expenditures of countries with a GDP total of [US $ '000]	1,075,127,808	Estimated domestic expenditures [US $ '000]				21,151

Group: high accessions and high GDP/c						
Countries	GDP at market prices	GDP/capita	Genebank accessions	Foreign aid received 1995	Foreign aid contributed 1995	Domestic expenditures 1995 incl. for.assist.
-	[cur. US $ '000]	[US $]	-		[US $ '000]	
Korea, Republic of	376,505,500	8,623	120,000	-	-	-
Spain	478,581,686	12,245	78,174	-	885	33,413
Israel	69,738,881	13,626	56,123	-	-	-
United Kingdom	941,423,919	16,274	114,495	-	17,531	70,154
Australia	289,390,395	16,553	94,768	-	-	-
Italy	991,385,682	17,149	80,000	-	-	27,208
Canada	546,349,515	19,907	212,061	-	3,580	1,584
Netherlands	309,226,799	20,373	67,374	-	-	-
Sweden	185,288,606	21,352	89,206	-	-	-
France	1,251,688,972	21,817	249,389	-	500	98,660
Germany	1,910,761,062	23,716	200,000	-	18,527	113,215
United States	6,259,899,105	24,509	550,000	-	-	20,433
Japan	4,190,470,930	33,671	202,581	-	-	6,480
Total [US $ '000]	17,800,711,053	249,815	2,114,171	-	41,023	371,146
13 Countries						
Foreign aid received by countries with GDP total of [US $ '000]	0	Estimated foreign aid received [US $ '000]				
Foreign aid contributed by countries with a GDP total of [US $ '000]	5,128,805,155	Estimated foreign aid contributed [US $ '000]			142,381	
Domestic Expenditures of countries with a GDP total of [US $ '000]	16,570,560,872	Estimated domestic expenditures [US $ '000]				398,699

Aggregated estimation						
Countries	GDP at market prices	GDP/capita	Genebank accessions	Foreign aid received 1995	Foreign aid contributed 1995	Domestic expenditures 1995 incl. for.assist.
-	[cur. US $ '000]	[US $]	-	[US $ '000]		
Total	23,828,725,133	-	-	17,359	50,111	475,045
165 countries			5,410,272	-	-	
Foreign aid received by countries with GDP total of [US $ '000]	2,689,068,976	Estimated foreign aid received [US $ '000]		268,413		
Foreign aid contributed by countries with a GDP total of [US $ '000]	5,905,900,650	Estimated foreign aid contributed [US $ '000]			164,288	
Domestic Expenditures of countries with a GDP total of [US $ '000]	19,444,682,461	Estimated domestic expenditures [US $ '000]				739,505

11 References

Agcaoili, Mercedita, Mark W. Rosegrant, 1995: Global and Regional Food Supply, Demand, and Trade Prospects to 2010. In: Nurul Islam (ed): Population and Food in the Early Twenty-First Century: Meeting Future Food Demand of an Increasing Population. International Food Police Research Institute, Washington, D.C., pp. 61-83.

AGRAR-Europe, 1997: Öffnung des Saatgutverkehrs kommt voran. In: Agrar-Europe, 6/97 of 10, February. P.: Länderbericht 30.

Alderman, Harold, Christina H. Paxson, 1992: Do the Poor Insure? A Synthesis of the Literature on Risk and Consumption in Developing Countries. Research Program in Development Studies, Discussion Paper 164. Princeton, New Jersey.

Alexandratos, Nikos (ed.), 1995: World Agriculture: Towards 2010: An FAO Study. John Wiley and Sons, New York.

Alika, J.E., M.E. Aken'Ova, C.A. Fatoukan, 1993: Variation among Maize (Zea mays L) accessions of Bendel State, Nigeria. Multivariate analysis of agronomic data. Euphytica 66: 65 -71.

Altieri, Miguel A, L.C. Merick, M.K. Anderson, 1987: Peasant agriculture and the conservation of crop and wild plant genetic resources. Conservation Biology 1:49-58.

Altieri, Miguel A., Camila Montecinos, 1993: Conserving Crop Genetic Resources in Latin America through Farmers' Participation. In: Christopher S. Potter, Joel I. Cohen, Dianne Janczewski (eds.): Perspectives on Biodiversity: Case Studies of Genetic Resource Conservation and Development. AAAS Publication, Washington, D.C. Pp.: 45-64.

André, H.M., M.I. Noti, P. Lebrun, 1994: The Soil Fauna: the other last Biotic Frontier. In: Biodiversity and Conservation 3, pp. 45-56.

Arrow, M., 1986: An assessment of the contingent valuation method. In: R.G. Cummings, D.S. Brookshire, W.D. Schulze (eds.): Valuing Environmental Goods. Rowman and Allanheld, Totowa, NJ.

Artuso, Anthony, 1994: Economic Analysis of Biodiversity as a Source of Pharmaceuticals. Paper presented at the Symposium on Biodiversity, Biotechnology and Sustainable Development at IICA Headquarters, San Jose, Costa Rica, April 12-14, 1994.

Arunachalam, V., 1986: Optimizing Conservation of Genetic Resources. In: NBPGR: Concepts and Prospects for Collection, Evaluation and Conservation. Summer Institute on Plant Genetic Resources, September 8-27, 1986. Reference Material. ICAR. New Delhi. Pp.: 101-112.

ASSINSEL (International Association of Plant Breeders), 1997. http://www.worldseed.org/~assinsel/assinselg.htm.

Auer, Marc, 1995: The German National Biodiversity Strategy. In: Kenton Miller, R., Steven M. Lanou: National Biodiversity Planning: Guidelines Based on Early Experiences Around the World. Pp.: 95-97. WRI, UNEP, IUCN. Washington, D.C.

Aylward, B.A., E.B. Barbier, 1992: Valuing Environmental Functions in Developing Countries. In: Biodiversity and Conservation, Vol. 1, pp: 34-50.

Banham, Will, 1993: Biodiversity: What is it? Where is it? How and why is it Threatened?. New Series Discussion Papers No. 33. Development and Project Planning Centre. University of Bradford.

Barbier, Edward B., Joanne C. Burgess, Carl Folke, 1994: Paradise Lost? The ecological economics of biodiversity. Earthscan, London.

Barbier, Edward B., W.M. Adams, K. Klimmage, 1991: Economic Valuation of Wetland Benefits: The Hadejia-Jama'are Floodplain, Nigeria. LEEC Discussion Paper 91-02, IIED, London.

Barhlott, Wilhelm, Wilhelm Lauer, Anja Placke, 1996: Global Distribution of Species Diversity in Vascular Plants: towards a World Map of Phytodiversity. In: Erdkunde, Band 50, 4, pp.317-327.

Barton, J., W. Siebeck, 1994: "Material transfer agreements in genetic resource exchange. The case of the International Agricultural Research Centres"; Issues in Genetic Resources, No. 1; IPGRI, Rome, May 1994.

BAZ (Bundesanstalt für Züchtungsforschung an Kulturpflanzen), 1997: Jahresbericht 1996. BAZ, Quedlinburg.

BDP (Bundesverband Deutscher Pflanzenzüchter e.V.), 1997a: Wer ist Kleinerzeuger? BDP, 15 NB 1/97. Bonn.

BDP, n.d.: Advantage. Multilateral System. Plant Genetic Resources from the point of view of German Plant Breeding. BDP, Bonn.

Berg, T., A. Bjornstad, C. Fowler, T. Skroppa, 1991: Technology Options and the Gene Struggle. NorAgric Occasional Paper, Norwegian Centre for International Agricultural Development, Agricultural University of Norway.

Binswanger, Hans P., V.W. Ruttan (eds.), 1978: Induced Innovation: technology, institutio and development. Baltimore-Londres, J. Hopkins University Press.

Bishop, R., 1978: Economics of a Safe Minimum Standard. In: American Jouranl of Agricultural Economics. 57, pp: 10-18.

BML (Bundesministerium für Ernährung, Landwirtschaft und Forsten), 1994: Statistisches Jahrbuch über Ernährung, Landwirtschaft und Forsten 1994.

BMU/CHM (German Clearing House Mechanism Secretariat), 1997: The German Clearing-House of the Convention on Biological Diversity. Discussion paper presented at the 2. SBSTTA Meeting in Montréal, 2.-6. September 1996.

Bolster, J.S., 1985: Selection for perceptual distinctiveness: evidence from Aguarana cultivars of Manihot esculenta. In: Economic Botany 39: 310-325.

Bommer, D., 1995: Personal Communication. Former Director General of FAL, Braunschweig.

Boongarts, John, 1996: Population Pressure and the Food Supply System in the Developing World. In: Population & Development Review. Vol. 22, No. 3, Sept. 1996, pp. 483-503.

Boserup, E., 1965: The condition of agricultural growth. Allen and Unwing, London.

Bowes, M.D., J.V. Krutilla, 1989: Multiple-Use Management: The Economics of Public Forestlands. Resources for the Future. Washington, DC.

Boyle, K.H., R.C. Bishop, 1986: The Economic Valuation of Endagered Species in Wildlife. Transactions of the Fifty-First North American Wildlife and Natural Resource Council.

Boyle, K.H., R.C. Bishop, 1987: Toward total Valuation of Great Lakes Fishery Resources. In: Water Resources Research, Vol. 5, pp: 943-950.

Braden, J.B., C.D. Kolstad (eds.), 1991: Measuring the Demand for Environmental Quality. Elsevier, New York, 1991.

von Braun, Joachim, 1994a: Genes and Biodiversity: new scarcities and rights challenge agricultural economics research. In: Quarterly Journal for International Agriculture. 4/1994. S. 345-348.

von Braun, Joachim, 1994b: Die langfristige Herausforderung der Ernährungssicherung: Politiklaternativen unter Bevölkerungsdruck.

von Braun, Joachim, Detlef Virchow, 1996: Economic Evaluation of Biotechnology and Biodiversity in Developing Countries. In: Agriculture and Rural Development. Vol. 3, No 1/1996.

von Braun, Joachim, Detlef Virchow, 1997: Conflict-Prone Formation of Markets for Genetic Resources: Institutional and Economic Implications for Developing Countries. Quarterly Journal for International Agriculture, 1/1997. Pp.: 6-38.

Brown, G.M., 1993: The economic value of elephants. In Edward B. Barbier (ed.): Economics and ecology: New frontiers and sustainable development. Chapman and Hall, London.

Brown, Gardner Jr., Jon H. Goldstein. 1984. A model for valuing endangered species. Journal of Environmental Economics and Management 11: 303-09.

Brown, Gardner Jr., 1990: Valuation of genetic resources. In: Gordon H. Orians, Gardner M. Brown Jr., William E. Kunin, Joseph E. Swierzbinski (eds.): The Preservation and Valuation of Biological Resources. Seattle: University of Washington Press, pp.: 203-28.

Brush, Stephen B, 1991: Farmer conservation of New World crops: the case of Andean potatoes, Diversity 7:75-79.

Brush, Stephen B, 1991: Farmer conservation of New World crops: the case of Andean potatoes, Diversity 7, pp.: 75-79.

Brush, Stephen B B., 1991: Farmer conservation of New World crops: the case of Andean potatoes. Diversity 7:75-79.

Brush, Stephen B., 1994: Providing Farmers Rights through the in situ conservation of crop genetic resources. Background study paper No. 2, Commission on Plant Genetic Resources. FAO, Rome.

Brush, Stephen B., Erika Meng, 1996: Farmers' Valuation and Conservation of Crop Genetic Resources. Paper presented at the CEIS - Tor Vergata University, Symposium on the Economics of Valuation and Conservation of Genetic Resources for Agriculture, 13-15 May, 1996.

BSA (Bundessortenamt), 1996: The Bundessortenamt. Hannover.

Bücken, Stefan, 1997: Personal communication at the genebank of the Federal Centre for Breeding Research on Cultivated Plants, Braunschweig.

BUKO (Bundeskongreß Entwicklungspolitischer Aktionsgruppen), 1996: Erhalt und Nutzung biologischer Vielfalt in der deutschen Landwirtschaft. In: Saatgut, Forschung und Züchtung in der biologisch-dynamischen Landwirtschaft. Dokumentation zur Ausstellung. Pp.: 24-26.

Busch, Lawrence, William B. Lacy, Jeffrey Burkhardt, 1989: Ethical and policy issues. In Biotic diversity and germplasm preservation, global imperatives. In: Lloyd Knutson, Allan K. Stoner (eds.): Invited papers presented at a symposium held May 9-11, 1988, at the Beltsville Agricultural Research Center, Beltsville, Maryland; Dordrecht: Kluwer Academic Publishers, pp.: 43-62.

Cabanilla, V.R., M.T. Jackson, T.R. Hargrove, 1993: Tracing the ancestry of rice varieties. 17th International Congress of Genetics. Volume of Abstracts, p112, 15-21 August 1993.

Cambolive-Piat, Maddy, 1996: Proposed Basel Action Plan. Presentation at the Industry workshop: The Conservation and Utilisation of Plant Genetic Resources for Food and Agriculture. February 15-16. Basel.

Cansier, Dieter, 1993: Umweltökonomie. G. Fischer, Stuttgart, Jena.

Carson, R.T. et al., 1992: A Contigent Valuation Study of Lost Passive Use Values Resulting from the Exxon Valdez OilSpill. Report to theAttorney General of the State of Alaska, Vol. 1, Natural Resource damage Assessment, Inc.

Carson, R.T. et al., 1995: A bibliography of contingent valuation studies and papers. Natural Resources Damage Assessment, Inc., La Jolla, Calif.

Ceccarelli S, Valkoiun J, Erskine W, Weigland S, Miller R & van Leur JAG., 1992: Plant genetic resources and plant improvement as tools to develop sustainable agriculture. Experimental Agriculture. 28, pp.: 89-98.

Ceccarelli, S., 1993: Specific adaptation and breeding for marginal conditions. Proceedings from the 18th EUCARPIA Fodder Crops Section Meeting, Loen, Norway, 25-28 August 1993.

CGIAR (Consultative Group on International Agricultural Research), 1996: The Financial Requirements of the 1996 CGIAR Research Agenda – MTM/95/05.

Chambers, R., 1994: Introduction. In: I. Scoones, J. Thompson (eds.): Beyond Farmer First: Rural Peoples Knowledge, Agricultural Research and Extension Practice. Intermediate Technology, London.

Chandel, K.P.S., 1996: Personal Communication. NBPGR, Director. New Delhi.

Chandel, K.P.S., R.S. Rana, 1995: Plant Genetic Resources System – an Indian Perspective. In: R.K. Arora, V. Ramanatha Rao (eds.): Proceedings of the South Asia National Coordinators Meeting on Plant Genetic Resources. Dhaka, Bangladesh, 10-12.

Chavez, R., W.M. Roca, J.T. Williams, 1987: IBPGR-CIAT collaborative project on a pilot in vitro active genebank. FAO/IBPGR PGR NL 71:11-13.

Ciriacy-Wantrup, S., 1952: Resource Conservation, Economics and Policies. Berkeley, University of California Press.

CITES (Convention on the International Trade in Endangered Species of Wild Fauna and Flora), 1993: Convention of International Trade in Endangered Species of Wild Fauna and Flora. CITES Secretariat, Geneva.

Clawson, M., J. Knetsch, 1966: Economics of Outdoor Recreation. John Hokins University Press, Baltimore.

Clegg, M.T., B.K. Epperson, A.H.D Brown, 1992: Genetic diversity and reproductives systems. In: Dattee et al. (eds.) Reproductive Biology and Plant Breeding. Springer, pp.: 311-324.

CMIE (Centre for Monitoring Indian Economy), 1988: Basic Statistics Relating to the Indian Economy. Vol. I: All-India. CMIE Pvt. Ltd.

Coase, R. H., 1960: The Problem of Social Cost. In: Journal of Law & Economics. 3/1960

Cooper, H. David, Elizabeth Cromwell, 1994: In situ conservation of crop genetic resources in developing countries: the influence of economic, policy and institutional factors. Draft Discussion Paper1. Rural Resources Management Group, Rural Poverty and Resources Research Programme. Funded by Natural Resources and Environment Department, ODA.

Cooper, H. David, Jan Engels, Emile Frison, 1994: A Multilateral System for Plant Genetic Resources: Imperatives, Achievements, and Challenges. Issues in Genetic Resources No. 2, May 1994.

Correa, Carlos, 1994: Sovereign and Property Rights over Plant Genetic Resources. Commission on Plant Genetic Resources. Background Study Paper No. 2. FAO, Rome.

Cowen, Tyler (Ed.), 1988: The Theory of Market Failure. A Critical Examination. George Mason University Press, Fairfax, Virginia.

Cromarty, A.S., R.H. Ellis, E.H. Roberts, 1985: The design of seed storage facilities for genetic conservation. IBPGR Handbooks for Genebanks No 1. IBPGR, Rome.

Cromwell, E., E. Friis-Hansen, M. Turner, 1992: The Seed Sector in Developing Countries. A framework for performance analysis. ODI Working Paper 65.

Dahl, K., G.P. Nabhan, 1992: From the grassroots up: The conservation of plant genetic resources by grassroots organizations – the "Latter-Day Noahs" of North America. Diversity 8 (2), pp 28-31.

Dalryample, D.G., 1986: Development and Spread of High-Yielding Wheat Varieties in Developing Countries., 7th ed. Washington DC: US Agency for International Development.

Davis, S.D., V.H. Heywood, A.C. Hamilton (Eds.), 1994: Centres of Plant Diversity. A Guide and Strategy for their Conservation. Vol. 1, WWF & IUCN, Cambridge.

DBV/BDP (Deutscher Bauernverband/Bundesverband Deutscher Pflanzenzüchter), 1996: Kooperationsvereinbarung des Deutschen Bauernverbandes e.V. und des Bundesverbandes Deutscher Pflanzenzüchter e.V., Bonn. Agreement signed by Constantin Freiherr Heereman, President of DBV and Dr. Wilhelm Graf von der Schulenburg, President of BDP. DBV/BDP, Bonn.

de Kathen, André, 1996: Gentechnik in Entwicklungslänern: Ein Überblick: Landwirtschaft. Umweltbundesamt, Berlin.

Dickson, Andrew, 1996: The Role of Industry Organisations in the Negotiation Process. Presentation at the Industry workshop: The Conservation and Utilisation of Plant Genetic Resources for Food and Agriculture. February 15-16. Basel.

Dobson, Andrew, Jon Paul Rodriguez, W. Mark Roberts, David Wilcove, 1997: Geographical Distribution of Endangered Species in the United States. In: Science, Vol. 275, 24 January 1997.

Dodds, J.H., Z. Huaman, R. Lizarraga, 1991: Potato germplasm conservation. In: In vitro Methods for Conservation of Plant Genetic Resources. J.H. Dodds (ed.) London, Chapman and Hall 93-109.

Ehrlich, P. R., E. O. Wilson, 1991: Biodiversity studies: science and policy. Science 253, pp.: 758-762.

Ellis, R.E., T. Hong, E.H. Roberts, 1990: An intermediate category of seed storage behaviour? Botany 41, pp.: 1167-1174.

Endres, Alfred, Brigitte Staiger, 1995: Umweltökonomie. In: Norbert Berthold (ed.): Allgemeine Wirtschaftstheorie. Neuere Entwicklungen. Pp: 75 - 87. Verlag Franz Vahlen, München.

Engelman, Robert, Pamela LeRoy, 1995: Conserving Land: Population and Sustainable Food Production. Population and Environment Program. Population Action International. Washington, D.C.

Engels, J.M.M., J.G. Hawkes, M. Worede, 1991: Plant Genetic Resources of Ethiopia. Cambridge University Press, Cambridge.

Esquinas-Alcázar, José, 1996: Farmers' Rights. Paper presented at the CEIS - Tor Vergata University, Symposium on the Economics of Valuation and Conservation of Genetic Resources for Agriculture, 13-15 May, 1996.

Evenson, Robert E., D. Gollin, 1996: Genetic resources, international organizations, and rice varietal improvement. Economic Development and Cultural Change, Yale University and University of Minnesota, USA.

Evenson, Robert E., 1993. Genetic Resources: Assessing Economic Value. Manuscript. New Haven, Conn.: Yale University, Department of Economics.

Evenson, Robert E., 1994: The Valuation of Crop Genetic Resources Preservation, Conservation and Use, Commission on Plant genetic Resources, FAO, (unpublished).

Evenson, Robert E., 1996a: The Valuation of Crop Genetic Resource Preservation, Conservation and Use. Manuscript prepared for the Secretariat of the Commission on Plant Genetic Resources, March 1996. Yale, USA.

Evenson, Robert E., 1996b: Valuing genetic resources for plant breeding: hedonic trait value, and breeding function methods. Paper presented at the CEIS - Tor Vergata University, Symposium on the Economics of Valuation and Conservation of Genetic Resources for Agriculture, 13-15 May, 1996.

Evenson, Robert E., C.C. David, 1993: Adjustment and Technology: The Case of Rice. OECD Development Centre series, Paris.

FAO (Food and Agriculture Organisation), 1989: Resolution 5/89, Farmers' Rights. Report of the Conference of FAO, 25th Session, Rome, 11-29 November, 1989. C89/REP. FAO, Rome.

FAO, 1991: FAO Food Balance Sheets, 1984-1986. FAO, Rome.

FAO, 1993: Ex situ storage of seeds, pollen and in vitro cultures of perennial woody plants. FAO Forestry Paper 113. FAO, Rome.

FAO, 1993a: Harvesting Nature's Diversity. FAO, Rome.

FAO, 1993b: International Undertaking on Plant Genetic Resources. Commission on Plant Genetic Resources. Fifth Session, Rome, 19 – 23 April, 1993. CPGR/93/Inf.2. FAO, Rome.

FAO, 1994: Seed Review 1989-1990. FAO, Rome.

FAO, 1995a: Report of the Sixth Session of the Commission on Plant Genetic Resources, 19-30 June 1995, Document CPGR-6/95 REP. FAO, Rome.

FAO, 1995b: Revision of the International Undertaking. Stage Three: Legal and Institutional Options. Commission on Plant Genetic Resources, Sixth Session, 19-30 June. FAO, Rome.

FAO, 1995c: Revision of the International Undertaking on Plant Genetic Resources. Analysis of some Technical, Economic and Legal Aspects for Consideration in Stage II. Access to Plant Genetic Resources and Farmers' Rights. CPGR-6/95/8 Supp. FAO, Rome.

FAO, 1995d: Survey of existing data on ex situ collections of plant genetic resources for food and agriculture. CPGR -6/95/8 Annex of the Sixth Session of the FAO Commission on Plant Genetic Resources, June 19-30, 1995. FAO, Rome.

FAO, 1996a: The State of the World's Plant Genetic Resources for Food and Agriculture. Background Documentation prepared for the International Technical Conference on Plant Genetic Resources, Leipzig, Germany, 17 - 23 June, 1996. FAO, Rome.

FAO, 1996b: Global Plan of Action for the Conservation and Sustainable Utilization of Plant Genetic Resources for Food and Agriculture. FAO, Rome.

FAO, 1996c: Report of the International Technical Conference on Plant Genetic Resources, Leipzig, Germany, 17-23 June 1996. Document: ITCPGR/96/Rep. FAO, Rome.

FAO, 1996d: Revision of the International Undertaking on Plant Genetic Resources. Issues for Consideration in Stage II: Access to Plant Genetic Resources, and Farmers' Rights. Commission on Plant Genetic Resources, 19-30 June. FAO, Rome.

FAO, 1996e: Rome Declaration on World Food Security and the Summit Plan of Action, adopted at the opening session on November 13. FAO, Rome.

FAO, 1996f: The Fourth International Technical Conference in the Context of the FAO Global System for the Conservation and Utilization of Plant Genetic Resources for Food and Agriculture. International Technical Conference on Plant Genetic Resources. ITCPGR/96/INF/2. FAO, Rome.

FAO, 1996g: Revision of the Cost Estimates of the Global Plan of Action. Commission on Genetic Resources for Food and Agriculture. Third Extraordinary Session, December 1996. CGRFA-EX3/96/INF. 1Annex. FAO, Rome.

FAO, 1996h: Respective Country Report. Submitted to FAO in the preparatory process for the International Technical Conference on Plant Genetic Resources, 1996.

FAO, 1997a: Background Documentation provided by the International Union for the Protection of New Varieties of Plants (UPOV). Commission on Genetic Resources for Food and Agriculture. CGRFA-7/97/Inf.5. Rome.

FAO, 1997b: Progress Report on the Global System for the Conservation and Sustainable Utilisation of Plant Genetic Resources for Food and Agriculture. Commission on Genetic Resources for Food and Agriculture. Seventh Session, 15-23 May. FAO, Rome.

FAO/IPGRI, 1994: Genebank Standards. Rome.

Fischbeck, G., 1981: The usefulness of gene banks - perspectives for the breeding of plants. In: UPOV Symposium: The use of Genetic Resources in the Plant Kingdom.

Fischbeck, G., 1992: Barley cultivar development in Europe. Success in the past and possible changes in the future. In: L. Munk (ed.): Barley Genetics VI. Vol. II. Pp.: 885-901. Munksgaard Intl. Publ. Ltd. Copenhagen.

Fisher, A.C., W.M. Hanemann, 1983: Option Value and the Extinction of Species. Working Paper 269, Giannini Foundation of Agricultural Economics. University of California, Berkeley.

Fowler, Cary, 1994: Unnatural Selection: Technology, Politics and Plant Evolution. Gordon and Breach Science Publishers, Yverdon.

Fowler, Cary, Pat Mooney, 1990: Shattering: Food, Politics, and the Loss of Genetic Diversity. University of Arizona Press, Tucson.

Frankel, O.H., J.J. Burdon, W.J. Peacock, 1995: Landraces in transit. The threat perceived. In: Diversity, 11, pp.: 14-15.

Freeman, A.M., 1991: Valuing Environmental Resources under Alternative Management Regimes. In: Ecological Economics, Vol. 3, pp: 247-256.

Frese, Lothar, 1996: Personal Communication. Braunschweig Genetic Resources Centre (BGRC), Head.

Frison, Emile A., Wanda W. Collins, Suzanne L. Sharrock, 1997: Global Programs: A New Vision in Agricultural Research. Issues in Agriculture 12. CGIAR, Washington, D.C.

GABG (German Association of Botanic Gardens), 1998: Contribution of the German Botanic Gardens to the Conservation of Biodiversity and Genetic Resources - Assessment and Development Concept. Homepage of GABG.

Gäde, Helmut, 1995: Personal Communication. Institute of Crop Science and Plant Breeding, Gatersleben. Head of regeneration program.

Garrod, G. K. Willis 1991: Some Empirical Estimates of Forest Amenity Value. Working Paper 13. Countryside Change Centre, University of Newcastle-upon-Tyne.

Gass, Thomas, 1996: Access to Plant Genetic Resources, Farmers' Rights and Equitable Sharing. Presentation at the Industry workshop: The Conservation and Utilisation of Plant Genetic Resources for Food and Agriculture. February 15-16. Basel.

GCA (Greenpeace Central America), 1996: Introduction and Use of Genetically Modified Organisms in Guatemala. A Greenpeace Report. Guatemala.

GEF (Global Environment Facility), 1997: Quarterly Operational Report. June 1997. GEF, Washington, D.C.

Gehl, Jens, 1997a: Kosten-Wirksamkeits-Analyse zur Bewahrung genetischer Ressourcen. Fallbeispiele aus Deutschland und Peru. Diplomarbeit. Rheinischen Friedrich-Wilhelms-Universität zu Bonn.

Gehl, Jens, 1997b: Personal communication in Kiel, December.

Glachant, M., 1991: La diversité biologique végétale: éléments d'économie. CERNA. Paris.

Gollin D, 1996: Valuing crop genetic resources for varietal improvement: the case of rice. Paper presented at the CEIS - Tor Vergata University, Symposium on the Economics of Valuation and Conservation of Genetic Resources for Agriculture, 13-15 May, 1996.

Goodman, M.M., 1985: ExoticMaize Germplasm: Status, Prospects and Remedies. Iowa State Jour. Res., 59. Pp.: 497-527.

Grohs, Florian, 1994: Economics of Soil Degradation, Erosion and Conservation: A Case Study of Zimbabwe. Kiel.

Groombridge, Brian (Ed.), 1992: Global Biodiversity. Status of the Earth's Living Resources. A Report compiled by the World Conservation Monitoring Centre. Chapman & Hall, London.

Gupta, Anil K., 1991: Why does poverty persist in regions of high biodiversity?: A case for indigenous property right system. WP No. 938. Indian Institute of Management, Ahmedabad.

Gupta, Anil K., 1996: Personal Communication. Indian Institute of Management. Ahmedabad.

Gustafson, J. Perry, quoted in Agricultural Research Magazine, June 1997.

Hagemann, R., 1985: Valuing Marine Mammal Populations: Benefit Valuations in a Multi-Species ecosystem. National Marine Fisheries Service, Southwest Fisheries Center, La Jolla, Calif.

Hamilton, Kirk, Ernst Lutz, 1996: Green National Accounts: Policy Uses and Empirical Experience. The World Bank, Washington, D.C.

Hammer, K., H. Gäde, H. Knüpffer, 1994: 50 Jahre Genbank Gatersleben – eine Übersicht. In: Vorträg für Pflanzenzüchtung, Heft 27, 1994. Pp.: 333-383.

Hammer, Karl, 1995: Personal Communication. Institute of Crop Science and Plant Breeding, Gatersleben. Head of Genebank.

Hampicke, Ulrich, 1991: Naturschutz-Ökonomie. UTB.

Hanley, N., 1989: Valuing Rural Recreation Benefits: An Empirical Comparison of two Approaches. In: Journal of Agricultural Economics, Vol. 40, pp: 361-374.

Hardon, J.J., B. Vosman, Th.J.L. van Hintum, 1994: Identifying Genetic Resources and Their Origin: The Capabilities and Limitations of Modern Biochemical and Legal Systems. Commission on Plant Genetic Resources. Background Study Paper No. 3. FAO, Rome.

Harlan, Jack R., 1976: The plants and animals that nourish man. Scientific American 235, pp.:89-97.

Harlan, Jack R., 1971: Agricultural origins: centers and noncenters. Science volume 21, No. 6. Pp: 468-474.

Hawkes, J.G., 1991: Summary of Final Discussion and Recommendations In: International Workshop on Dynamic In-Situ Conservation of Wild Relatives of Major Cultivated Plants, Israel. In: Journal of Botany, 40, pp: 529-536.

Hayami, Yujiro, Vernon W. Rutan, 1985: Agricultural Development: An International Perspective. Johns Hopkins University Press, Baltimore.

Hazell, Peter, James L. Garrett, 1996: Reducing Poverty and Protecting the Environment: the Overlooked Potential of Less-Favored Lands. 2020 Brief 39, IFPRI, Washington, D.C.

Heidhues, Franz, 1994: Probleme internationaler Arbeitsteilung in der Agrarwirtschaft. In: Agrarwirtschaft 43, 1994, Heft 4/5. S. 191 - 197.

Heywood VH (ed.), 1995: Global Biodiversity Assessment. UNEP, Cambridge University Press, UK.

Hinegardner, R., 1976: Evolution of genome size. In: F.J. Ayala (ed.): Molecular Evolution. Sinauer Associates, Sunderland, Mass. Pp.: 179-199.

Hobbelink, Henk, 1989: Gene bank or gene morgue? In: Seedling, Vol. 6, No 5, p.:2.

Hodgson, G., J.A. Dixon, 1988: Logging versus Fisheries and Tourism in Palawan. Paer No. 7, East West Center. Honolulu, HI.

Hohl, A., C. Tisdell, 1993: How useful are Environmental Safety Standards in Economics? The Example of Safe Minimum Standards for Protection of Species. In: Biodiversity and Conservation. 2, pp: 168-181.

Holden, J.H.W., J.T. Williams (eds.), 1984: Crop Genetic Resources: Conservation and Evaluation. George Allen and Unwin, London.

Hossain, M., 1994: Recent developments in Asian rice economy: challenges for rice research. In: Workshop on Rice Research Prioritization in Asia, 15-22 Feb, 1994, IRRI, Los Banos, Phillippines.

Hoyt, E., 1988: Conserving the Wild Relatives of Crops. Rome: International Board for Plant Genetic Resources.

Hyde, Charles E., James A. Vercammen, 1997: Costly Yield Verification, Moral Hazard, and Crop Insurance Contract Form. In: Journal of Agricultural Economics, 48, (3), pp.: 393-407.

ICR (Indian Country Report), 1995: Country Report on Status of Plant Genetic Resources India. Submitted to FAO in the preparatory process for the International Technical Conference on Plant Genetic Resources, 1995.

ICRISAT (International Crops Research Institute for the Semi-Arid Tropics), n.d.: ICRISAT Standard Order Form.

Iltis, H.H., J.F. Doebley, R. Guzman, B. Pazy, 1979: Zea diploperennis (Graminae): A new teosinte from Mexico. Science 203:186-188.

IPGRI (International Plant Genetic Resources Institute), 1995: Geneflow. A Publication about the Earth's Plant Genetic Resources. IPGRI, Rome.

IPGRI, 1996: Access to plant genetic resources and the equitable sharing of benefits: a contribution to the debate on systems for the exchange of germplasm. Issues in Genetic Resources No. 4. IPGRI, Rome.

Iwanga, M., 1993: Enhancing links between germplasm conservation and use in a changing world. International Crop Science. I, pp: 407-413

Jain, H.K., 1992: GATT Proposals and India's Seed Policy. In: Seed Research Vol. 20 (2), pp.: 65-69.

Jenkins, Martin, 1992: Species Extinction. In Groombridge Pp:192-205.

Juma, Calestro, 1989: The Gene Hunters: Biotechnology and the Scramble for Seeds. Zed Books, London.

King, Ken, 1994: Incremental Cost as an Input to Decisions About the Global Environment. The Global Environment Facility (GEF), Washington, D.C.

Kloppenburg, Jack R., Daniel Lee Kleinman, 1987: Plant germplasm controversy - analyzing empirically the distribution of the world's plant genetic resources. Bioscience 37:190-198.

Kloppenburg, Jack R., Daniel Lee Kleinman, 1988: Seeds of Controversy: National Property versus Common Heritage. In: Jack R. Kloppenburg (ed.): Seeds and Sovereignty. The Use and Control of Plant Genetic Resources. Duke University Press, London. Pp.: 173-203.

Knoll, A.H., 1984: Patterns of Extinction in the Fossil Record of Vascular Plants. In: M.H. Nitecki (ed.): Extinctions. University of Chicago Press, Chicago, IL. Pp. 22-68.

Kuo, C.G., 1991: Conservation and distribution of sweet potato germplasm. In: J.H. Dodds (ed.): In vitro Methods for Conservation of Plant Genetic Resources. Chapman and Hall, London. Pp.: 123-148.

Kush, Anil, 1996: Personal Communication. Indo-American Hybrid Seeds, Director Biotechnology. Bangalore.

Kush, G.S., 1996: Personal communication in New Delhi, IRRI.

Lampietti, J. A., J. A. Dixon, 1995: To See the Forest for the Trees: A Guide to Non-Timber Forest Benefits. Environment Department Paper 13. The World Bank, Washington, D.C.

LeBuanec, Bernard, 1996: Conservation and Utilisation of Plant Genetic Resources for Food and Agriculture. Key issues for the seed and plant biotechnology industries. Presentation at the Industry workshop: The Conservation and Utilisation of Plant Genetic Resources for Food and Agriculture. February 15-16. Basel.

LEI-DLO (Agricultural Economic Institute), 1994: The World Seed Market. Rabobank Nederland and Ministry of Agriculture, Nature Management and Fisheries.

Leskien, Dan, Michael Flitner, 1997: Intellectual Property Rights and Plant Genetic Resources: Options for a Sui Generis System. Issues in Genetic Resources No. 6. IPGRI, Rome.

Lesser, William, 1994: Attributes of an Intellectual Property Rights System for Landraces. In: M.S. Swaminathan, , Vineeta Hoon (eds.): Farmers' and Breeders' rights. Background papers, CRSARD, Madras. Pp.: 233-250.

Lipton M., R. Longhurst, 1989: New Seeds and Poor People. Unwin Hyman, London.

Loomis, J., C. Sorg, D. Donnelly, 1986: Economic Losses to Recreational Fisheries due to Small Head Hydro Development. In: Journal of Environmental Management, Vl. 22, pp: 85-94.

Lutz, Ernst (ed.), 1993: Toward Improved Accounting for the Environment. An UNSTAT-World Bank Symposium. The World Bank, Washington, D.C.

Malyshev, L.I., 1975: The Quantitative Analysis of Flora: Spatial Diversity, Level of Specific richness, and Representativity of Sampling Areas. In: Botanicheskiy Zhurn. 60, pp. 1537-1550.

McNeely, Jeffrey A., 1996: The Conservation and Sustainable Use of Plant Genetic Resources. Presentation at the Industry workshop: The Conservation and Utilisation of Plant Genetic Resources for Food and Agriculture. February 15-16. Basel.

McNeely, Jeffrey A., Kenton R. Miller, Walter V. Reid, Russell A. Mittermeier, Timothy B. Werner. 1990. Conserving the world's biological diversity. Gland, Switzerland, and Washington, DC: Prepared and published by the International Union for Conservation of Nature and Natural Resources, World Resources Institute, Conservation International, World Wildlife Fund-US, and the World Bank.

Melaurie, B., O. Pungu, R. Dumont, M-F. Trouslot, 1993: The creation of an in vitro germplasm collection of yam (Dioscorea spp.) for genetic resources preservation. Euphytica 65:113-122.

Meng, Erika, J. Edward Taylor, Stephen Brush, 1997: Household Varietal Choice Decisions and Policy Implications for the Conservation of Wheat Landraces in Turkey. Draft. Paper presented at the Symposium: Building the Theoretical and Empirical Basis for the Economics of Genetic Diversity and Genetic Resources Conservation in Crop Plants. Held at Stanford University, Palo Alto, California, August, 17-19, 1997.

Mishra, Shri Chaturanan, 1996: Inauguration Speech of the Union Minister of Agriculture, 2nd International Crop Science Congress, New Delhi.

Mitchell, Donald O., Merlinda D. Ingco, 1995: Global and Regional Food Demand and Supply Prospects. In: Nurul Islam (ed): Population and Food in the Early Twenty-First Century: Meeting Future Food Demand of an Increasing Population. International Food Police Research Institute, Washington, D.C., pp. 49-60.

Mooney, Pat R., 1993: Exploiting local knowledge: international policy implications. In: W. de Boef, K. Amanor, K. Wellard, A. Bebbington, 1993: Cultivating Knowledge: genetic diversity, farmer experimentation and crop research. IT Publications, London.

Morris, Michael L., Paul W. Heisey, 1997: Achieving desirable levels of crop genetic diversity in farmers' fields: Factors affecting the production and use of improved seed. Paper presented at the Symposium: Building the Theoretical and Empirical Basis for the Economics of Genetic Diversity and Genetic Resources Conservation in Crop Plants. Held at Stanford University, Palo Alto, California, August, 17-19, 1997.

Munasinghe, Mohan, Ernst Lutz, 1993: Environmental Economics and Valuation in Development Decisionmaking. Environment Working Paper No. 51, The World Bank, Washington, DC.

Narayanan, 1996: Personal Communication. Southern Petrochemical Industries Corporation Ltd. Chennai (Madras).

Norton-Griffiths, M., C. Southey, 1994: The Opportunity Costs of Biodiversity Conservation in Kenya. In: Ecological Economics, 12, 1995, pp. 125-139.

Novak, F.J., 1990: Allium tissue cultures. In: J.L. Rabinocitch, J.L. Brewster (eds.): Onion and Allied Crops. CRC Press Inc., Florida. Pp.: 233-250.

NRC (National Research Council), 1972: Genetic Vulnerability of Major Crops, National Academy of Sciences: Washington D.C.

NRC (National Research Council), 1993: Managing Global Genetic Resources: Agricultural Crop Issues and Policies. National Academy Press, Washington, USA.

ODI (Overseas Development Institute), 1993: Briefing Paper, November 1993.

OECD (Organisation for Economic Cooperation and Development), 1996a: Intellectual Property, Technology Transfer and Genetic Resources: An OECD survey of current practices and policies. OECD, Paris.

OECD, 1996b: Saving Biological Diversity. Economic Incentives. OECD, Paris.

Oetmann, Anja, 1996: Expenditures for the Conservation and Utilisation of PGRFA in Germany. Data prepared for the Preparatory Process of the 4th International Technical Conference for Plant Genetic Resources. Unpublished.

Oldfield, Margery L. 1989: The value of conserving genetic resources. Sinauer Associates, Inc. Sunderland, Massachusetts

Paroda, R.S., 1986: Plant Genetic Resources and their Management. An Overview. In: NBPGR: Concepts and Prospects for Collection, Evaluation and Conservation. Summer Institute on Plant Genetic Resources, September 8-27, 1986. Reference Material. ICAR. New Delhi. Pp.: 1-18.

Paroda, R.S., 1996: India's National Bureau of Plant Genetic Resources Leads Nationwide System. In: Diversity, Vol. 12, No. 3. Pp.: 43-47.

Paroda, R.S., R.K. Aroba, 1986: Plant Genetic Resources – An Indian Perspective. NBPGR Sci. Monogr., 10, ICAR. NBPGR, New Delhi.

Pearce, D.W., E.B. Barbier, A. Markandya, 1990: Sustainable Development. Earthscan, London.

Pearce, D.W., E.B. Barbier, A. Markandya, S. Barrett, R.K. Turner, T. Swanson, 1991: Blueprint 2. Earthscan, London.

Pearce, David W., Rafaello Cervigni, 1994: The valuation of the contribution of plant genetic resources. In: Timothy M. Swanson et al. (eds.): The appropriation of the benefits of plant genetic resources for agriculture: An economic analysis of the alternative mechanisms for biodiversity conservation. FAO, Commission on Plant Genetic Resources. Background Study Paper No. 1. Presented at First Extraordinary Session, Rome, 7-11 November 1994. FAO, Rome.

Peeters, J.P., J.T. Williams, 1984: Toward better use of genebanks with special reference to information. Plant Genetic Resources Newsletter 60:22-32.

Perrings, Charles, David Pearce, 1994: Threshold Effects and Incentives for the Conservation of Biodiversity. In: Environmental and Resource Economics, 4. Pp.: 13-28.

Pimentel, David, C. Harvey, P. Resosudarmo, K. Sinclair, D. Kurz, M. McNair, S. Crist, L. Shpritz, L. Fitton, R. Saffouri, R. Blair, 1994: Environmental and Economic Costs of Soil Erosion and Conservation Benefits. College of Agriculture and Life Sciences, Cornell University, Ithaca, NY. Draft.

Plucknett, D.L., N.J.H. Smith, J.T. Williams, N.M. Anishetty, 1987: Gene Banks and the World's Food. University Press: Princetown, New Jersey.

Porceddu E., G. Jenkins (eds.), 1991: Seed Regeneration of Cross-Pollinated Species. AA Balkema, Rotterdam.

Pretty, Jules N., 1995: Regenerating Agriculture: Policies and Practice for Sustainability and Self-Reliance. Earthscan Publications Ltd., London.

Pundis, R.P.S., 1996: Personal Communication. ICRISAT.

Ragot, M., D.A. Hoisington, 1993: Molecular markers for plant breeding: comparisons of RFLP and RAPD genotyping costs. Theor. Appl. Genet. 86: 975-984.

Rana, R.S., K.P.S. Chandel, 1992: Crop Genetic Resources. In: T.N. Khoshoo, Manju Sharma (eds.): Sustainable Management of Natural Resources. National Academy of Sciences, India. Malhotra Publishin House, New Delhi. Pp.: 159-198.

Rana, R.S., K.P.S. Chandel, 1992: Crop Genetic Resources. In: T.N. Khoshoo, Manju Sharma (eds.): Sustainable Management of Natural Resources. National Academy of Sciences, India. Malhotra Publishin House, New Delhi. Pp.: 159-198.

Rani, M. Geetha, 1996: Personal communication. M.S. Swaminathan Research Foundation, Madras. Genebank Manager. November 1996.

Reid, Walter V., 1992: How many species will there be? In: T.C. Withmore, J.A. Sayer (eds.):Tropicla Deforestation and Species Extinction. Pp.: 55-74. Chapman and Hall, London.

Reid, Walter V., 1995: Gene Co-Ops and the Biotrade: Translating Genetic Resource Rights into Sustainable Development. Final Draft. Journal of Ethnopharmacology. World Resources Institute (WRI), Washington, D.C.

Reid, Walter V., Sarah A. Laird, Carrie A. Meyer, Rodrigo Gámez, Ana Sittenfeld, Daniel H. Janzen, Michael A. Gollin, Calestous Juma, 1993: Biodiversity Prospecting: Using Genetic Resources for Sustainable Development. World Resources Institute (WRI), USA, Instituto Nacional de Biodiversidad (INBio), Costa Rica, Rainforest Alliance, USA, African Centre for Technology Studies (ACTS), Kenya.

Repetto, Robert, William Magrath, Michael Wells, Christine Beer, Fabrizio Rossini, 1989: Wasting Assets: Natural Resources in the National Income Accountss. World Resources Institute, Washington.

Rhoades, Rober E., 1996: Stakeholder Analysis of the Global-Local Nexus in Plant Genetic Resources. Paper presented at the 1996 Annual Meeting of the Society for Applied Anthropology (SAA): Local-Global (Dis)articulaitons in Plant Genetic Resources Conservation.

Riley, R., 1989: Plant biotechnologies in developing countries: The plant breeders perspective In: A. Sasson, V. Costarini (eds.): Plant Biotechnologies for Developing Countries. Proceedings of an International Symposium organized by CTA and FAO, 26-30 June, 1989, Luxembourg.

Roberts, E.H., 1973: Predicting the storage life of seeds. Seed Science and Technology, 1: 499-514.

Roberts, R.A.J., W.J.A. Dick, 1991: Strategies for crop insurance planning. FAO Agricultural Services Bulletin No 86, FAO, Rome.

Robertson, J., 1992: Biosphere reserves: Relations with natural World Heritage sites. Parks 3:29-34.

Rosegrant, Mark W., Claudia Ringler, Roberta V. Gerpacio, 1997: Water and Land Resources and Global Food Supply. Plenary Paper prepared for the XXIII International Conference of Agricultural Economists on 'Food Security, Diversification, and Resource Management: Refocusing the Role of Agriculture?' Sacramento, California, August 10-16, 1997.

Rubin, J., G. Helfand, J. Loomis, 1991: A benefit-cost analysis of the northern spotted owl. In: Journal of Forestry, 89.

Ruitenbeek, H.J., 1992: Mangrove Management: An Economic Analysis of Management Options with a Focus on Bintuni Bay, Irian Java. Environmental Management Development in Indonesia Project. Environmental Reports, No. 8.

Salhuana, Wilfredo, Stephen Smith, 1996: Maize Breeding and Genetic Resources. Paper presented at the CEIS - Tor Vergata University, Symposium on the Economics of Valuation and Conservation of Genetic Resources for Agriculture, 13-15 May, 1996.

Sarbhoy, A.K., 1996: Personal Communication. Centre of Biosystematics, IARI. Chief Project-Coordinator. New Delhi.

Schäfer-Menuhr, A, 1996: Refinement of Cryopreservation Techniques for Potato. Braunschweig.

Schaffer, M, 1987: Minimum viable populations: coping with Uncertainty. In: M.E. Soulé (Ed.): Viable Populations for Conservation. Cambridge University Press, Cambridge, New York. Pp. 70-86.

Sehgal, S., 1996: Intellectual Property Rights and the seed industry in India. In: Diversity, Vol. 12, No. 3. Pp.: 79-80.

Semon S, 1995: The impact of the UPOV 1991 Act upon Seed Production and Research. Ms.C. Thesis, University of Edinburgh, Scotland.

Sen, Amartya K., 1981. Poverty and Famines: An Essay on Entitlement and Deprivation. Oxford: Clarendon Press.

Senrayan, R., 1996: Personal Communication. E.I.D. Parry (India) Ltd. Manager – Development. Chennai (Madras).

SGRP (System Wide Programme on Plant Genetic Resources) , 1996: Report of External Review Panel of the CGIAR Genebank Operations: CGIAR, Washington.

Shands, Henry, 1994: The National Plant Germplasm System: An Update. Presented at the Economic Research Service and the Farm Foundation Conference: Global Environmental Change and Agriculture: Assessing the Impacts, November 2, 1994. Washington, D.C.

Shands, Henry, 1996: Personal Communication. Associate Deputy Administrator for Genetic Resources. Agricultural Research Service. Department of Agriculture, Washington.

Sharma, S.P., 1996: Personal Communication. Division of Seed Science & Technology, IARC. Head. New Delhi.

Shiva, Vandana, et al., eds. 1991. Biodiversity: Social and ecological perspectives. London and New Jersey: Zed Books Ltd, with Penang, Malaysia: World Rainforest Movement.

Siddiq, E.A., 1996: Letter Communication from January 9, 1996. Indian Council of Agricultural Research. Deputy Director General.

Signor, P.W., 1990: The geological history of diversity. In: Annual Review of Ecology ans Systematics. 21, pp.: 509-539.

Simpson, D., R.A. Sedjo, J.W. Reid, 1996: Valuing biodiversity for use in pharmaceutical research. International Journal of Political Economy 104:163-185.

Simpson, R. David, Roger A. Sedjo, 1996: The Value of Genetic Resources for Use in Agricultural Improvement. Paper presented at the CEIS - Tor Vergata University, Symposium on the Economics of Valuation and Conservation of Genetic Resources for Agriculture, 13-15 May, 1996.

Singh, R.P., Suresh Pal, Michael Morris, 1995: Maize Research and Development and Seed Production in India: Contributions of the Public and Private Sectors. CIMMYT Economics Working Paper 95-03. CIMMYT, Mexico, D.F.

Smale, Melinda, 1996a: Global Trends in Wheat Genetic Diversity and International Flows of Wheat Genetic Resources. Part I of: CIMMYT World Wheat Facts and Trends, 1996. CIMMYT, Mexico, D.F.

Smale, Melinda, 1996b: Indicators of Genetic Diversity in Bread Wheats: Selected Evidence on Cultivars Grown in Developing Countries. Paper presented at the CEIS - Tor Vergata University, Symposium on the Economics of Valuation and Conservation of Genetic Resources for Agriculture, 13-15 May, 1996.

Smale, Melinda, 1997: The Green Revolution and Wheat Genetic Diversity: Some unfounded Assumptions. In: World Development, Vol. 25, No. 8, pp: 1257-1269. Elsevier Sciences Ltd. Great Britain.

Smith, Stephen, Wilfredo Salhuana, 1996: The Role of Industry in the Conservation and Utilization of Plant Genetic Resources. Presentation at the Industry workshop: The Conservation and Utilisation of Plant Genetic Resources for Food and Agriculture. February 15-16. Basel.

Stanwood, P.C., L.N. Bass, 1981: Seed germplasm preservation using liquid nitrogen (-196°C), Seed Technology 5:26-31.

Stoll, J.R., L.A. Johnson, 1984: Concepts of Value, Nonmarket Valuation and the Case of the Whooping Crane. In: Transactions of the Forty-Ninth North American Wildlife and Natural Resources Conference, Vol. 49, pp: 382-393.

Stork, N., 1993: How many Species are there? In: Biodiversity and Conservation 2, pp.: 215-232.

Svarstad, Hanne, 1994: National Sovereignty and genetic resources. In: Vincente Sanchez, Calestous Juma (eds.): Biodiplomacy – Genetic Resources and International Relations. ACTS Press, Nairobi. Pp.: 46-65.

Swaminathan, M.S. (ed.), 1996a: Farmers' Rights and Plant Genetic Resources. Macmillan India Ltd. Madras.

Swaminathan, M.S. (ed.), 1996b: Agrobiodiversity and Farmers' Rights: Proceedings of a Technical Consultation on an Implementation Framework for Farmers' Rights. M.S. Swaminathan Research Foundation, Madras.

Swaminathan, M.S., 1996c: Compensating Farmers and Communities through a Global Fund for Biodiversity Conservation for Sustainable Food Security. In: Diversity A News Journal for the International Genetic Resources Community. Vol. 12, No.3. Pp.: 73-75.

Swaminathan, M.S., 1996d: Personal Communication. M.S. Swaminathan Research Foundation, Madras.

Swanson, T.M., D.W. Pearce, R. Cervigni, 1994: The appropriation of the benefits of plant genetic resources for agriculture: An economic analysis of the alternative mechanisms for biodiversity conservation. FAO Commission on Plant Genetic Resources, Background Study Paper No. 1. FAO, Rome.

Swanson, Timothy, 1996: Development, Agriculture and Diversity: Externalities in the Diffusion of Agriculture. Paper presented at the CEIS - Tor Vergata University, Symposium on the Economics of Valuation and Conservation of Genetic Resources for Agriculture, 13-15 May, 1996.

Tapia, M.E., A. Rosas, 1993: Seed fairs in the Andes: a strategy for local conservation of plant genetic resources. In: W. de Boef, K. Amanor, K. Wellard (eds.): Cultivating Knowledge. Genetic diversity, farmer experimentation and crop research. Pp. 111-118. Intermediate Technology Publications.

Tarp, 1995: Personal communication at FAO.

The Indian Express, 1996: The Indian Express, New Delhi, November 20, p.: 11.

Thiele, Rainer (1994): Zur ökologischen Bewertung tropischer Regenwälder. In: Die Weltwirtschaft. Heft 3, 1994. S. 363-378. Institut für Weltwirtschaft (IfW), Kiel.

Trommeter, Michel, 1993: Rationalisation économique de la conservation des ressources génétiques végétales. Université Pierre Mendes France, UFR de Sciences Economiques de Grenoble.

Turner, R.K. (ed.), 1993: Sustainable Environmental Economics and Management. Belhaven Press, London.

UN (United Nations), 1993:Handbook on Integrated Environmental and Economic Accounting.

UN, 1995: Statistical Yearbook 1993, 40th Issue, New York.

UNEP (United Nations Environment Programme), 1994a:Convention on Biological Diversity. Text and Annexes. Interim Secretariat for the Convention on Biological Diversity, Geneva Executive Center, Switzerland.

UNEP, 1994b: Policy, strategy, programme priorities and eligibility criteria for access to and utilization of financial resources, UNEP/CBD/COP/1/17, Annex I.

UPOV (Union pour la Protection des Obtentions Végétales), 1992: International Convention for the Protection of New Varieties of Plants of December 2, 1996, as Revised at Geneva on November 10, 1972, on October 23, 1978, and on March 19, 1991. UPOV, Geneva.

UPOV, 1995: Internationaler Verband zum Schutz von Pflanzenzüchtungen. Allgemeine Infromationsbroschüre. UPOV, Genf.

Van Laecke, K., E. Van Bockstaele, M. De Loose, 1995: Evaluation of RAPD-markers for the identification of ryegrass varieties. UPOV Working Group on Biochemical and Molecular Techniques and DNA Profiling in Particular, UPOV Paper BMT/3/3.

van Zanten, Jasper, 1996:Technology Transfer including Biosafety. Presentation at the Industry workshop: The Conservation and Utilisation of Plant Genetic Resources for Food and Agriculture. February 15-16. Basel.

Vaughan, D.C., T. Chang, 1992: In situ conservation of rice genetic resources. In: Economic Botany, 46 (4), pp.: 368-383.

Vavilov, Nicolai I., 1926: Studies on the origin of cultivated plants. In: Bulletin of Applied Botany, Genetics, and Plant Breeding, 16, pp.: 139-246.

Vellvé, R., 1992: Saving the seed. Genetic diversity and European agriculture. Earthscan Publications Ltd, London.

Vicente, P.R., 1994: The MASIPAG program: an integrated approach to genetic conservation and use. In: Growing Diversity in Farmers Fields. Proceedings of a Regional Seminar for Nordic Development Cooperation Agencies, Lidingo, Sweden, 26-28 September, 1993.

Virchow, Detlef, 1996a: National and International Expenditures for Conservation and Utilisation of Plant Genetic Resources for Food and Agriculture in 1995. A draft background paper for the document ITCPGR/96/INF/1: Current Expenditures for the Conservation and Utilisation of Plant Genetic Resources for Food and Agriculture. Prepared for the International Technical Conference on Plant Genetic Resources, Leipzig, 17 - 23 June, 1996. FAO, Rome.

Virchow, Detlef, 1996b: Costing the Global Plan of Action for the Conservation and Sustainable Utilisation of Plant Genetic Resources for Food and Agriculture. A draft background paper for the document ITCPGR/96/5: The Global Plan of Action for Conservation and Utilisation of Plant Genetic Resources for Food and Agriculture. Prepared for the International Technical Conference on Plant Genetic Resources, Leipzig, 17 - 23 June, 1996.

Virchow, Detlef, 1997a: How to optimise Plant Genetic Resources Conservation: Cost-Efficiency of Different Property Regimes. Background Paper prepared for a Poster Demonstration at the XXIII International Conference of Agricultural Economists on 'Food Security, Diversification, and Resource Management: Refocusing the Role of Agriculture?' Sacramento, California, August 10-16, 1997.

Virchow, Detlef, 1997b: Who is Bearing the Costs of Plant Genetic Resources Conservation? CIMMYT Symposium: Building the Theoretical and Empirical Basis for the Economics of Genetic Diversity and Genetic Resource Conservation in Crop Plants. Stanford University, California. August, 1997.

Vogel, Joseph Henry, 1994: Genes for Sale. Privatization as a Conservation Policy. Oxford University Press, Oxford.

WCMC (World Conservation Monitoring Centre), Faculty of Economics, Cambridge University, 1996: Industrial Reliance upon Biodiversity. WCMC, Cambridge.

Weltzien-Rattunde, Eva, 1996: Personal Communication. ICRISAT, Senior Scientist, Breeding.

Whitehead, J.C., 1993: Total Economic Values for Coastal and Marine Wildlife: Specification, Validity, and Valuation Issues. In: Marine Resource Economics, Vol. 8, pp: 119-132.

WIEWS, 1996: Datainformation from the World Information and Early Warning System, FAO

Williamson, O.E., 1975: Markets and Hierarchies. Analysis and Antitrust Implications. New York.

Wilson, E.O. (ed.), 1988: Biodiversity. Washington, DC: National Academy Press.

Wilson, E.O., 1992: The Diversity of Life. Penguin, London.

Withers, L.A., 1990: Cryopreservation of plant cells. Biol. J. Linnean Society 43: 31-42.

Wolf, E.C., 1987: Beyond the Green Revolution: New Approaches for Third World Agriculture, Worldwatch paper 1987/1, Worldwatch Institute, Washington, D.C.

Wolfe, M, 1992: Barley diseases: maintaining the value of our varieties. Barley Genetics VI, vol. II, pp. 1055-1067.

Wood, D., 1988a: Crop germplasm: common heritage or farmers heritage? In: Jack R. Kloppenburg (ed.) Seeds and Sovereignty, Duke University Press.

Wood, D., 1988b: Introduced crops in developing countries - a sustainable agriculture? Food Policy , May 1988, p167-177 .

Wood, D., J. Lenne, 1993: Dcnamic management of domesticated biodiversity by farming communities. In: Proceedings of the UNEP/Norway Expert Conference on Biodiversity, Trondheim, Norway.

Worede, M., 1992: The role of Ethiopian farmers in the conservation and utilization of crop genetic resources. First Int. Crop Sci. Congress, Ames, Iowa.

World Bank, 1997: Expanding the Measure of Wealth: Indicators of Environmentally Sustainable Development. Environmentally Sustainable Development Studies and Monographs Series, No. 17. The World Bank, Washington.

Wright, B.D., 1996: Intellectual property and farmers' rights. Paper presented at the CEIS - Tor Vergata University, Symposium on the Economics of Valuation and Conservation of Genetic Resources for Agriculture, 13-15 May, 1996.

Wright, Brian D. 1995. Agricultural genetic resource policy: Towards a research agenda. Paper prepared for presentation at the Technical Consultation on Economic and Policy Research for Genetic Resource Conservation and Use, International Food Policy Research Institute, Washington, DC, June 21-22, 1995. Manuscript. University of California at Berkeley, Department of Agricultural and Resource Economics.

Wych, R.D., D.C.Rasmusson, 1983: Genetic improvement of malting barley cultivars since 1920. Crop Science. 23:1037-1040.

ZADI (Zentralstelle für Agrardokumentation und –information), 1995: Jahresbericht 1994. ZADI, Bonn.

Zedan, Hamdallah, 1996: The Role of UN Organisations in the Negotiation Process. Presentation at the Industry workshop: The Conservation and Utilisation of Plant Genetic Resources for Food and Agriculture. February 15-16. Basel.

Zeven, A.C., P.M. Zhukovsky, 1975: Dictionary of Cultivated Plants and their Centres of Diversity. Centre for Agricultural Publishing and Documentation: Wageningen.

Zilberman, David, Cherisa Yarkin, Amir Heiman, 1997: Agricultural Biotechnology: Economic and International Implications. Paper presented at the Symposium: Building the Theoretical and Empirical Basis for the Economics of Genetic Diversity and Genetic Resources Conservation in Crop Plants. Held at Stanford University, Palo Alto, California, August, 17-19, 1997.

Index

Printing: Weihert-Druck GmbH, Darmstadt
Binding: Buchbinderei Schäffer, Grünstadt

Springer and the environment

At Springer we firmly believe that an international science publisher has a special obligation to the environment, and our corporate policies consistently reflect this conviction.

We also expect our business partners – paper mills, printers, packaging manufacturers, etc. – to commit themselves to using materials and production processes that do not harm the environment. The paper in this book is made from low- or no-chlorine pulp and is acid free, in conformance with international standards for paper permanency.